What Do Bees Think About?

ANIMAL WORLDS

Jessica Serra, Series Editor

What if,
instead of looking at animals
through our own eyes,
we looked through theirs?

Recent scientific discoveries offer us a new perspective on the animal kingdom, shattering the myth that once equated animal behavior with that of machines. We now know that humans are not the only beings with intelligence, emotions, and language skills.

Even though animals share our environment, they perceive and understand it in their own way. Equipped with specific sensory equipment, they selectively pick up certain meaningful signals and evolve in a world of their own. This means that our human world is only one among millions of others.

Shifting our perspective to reflect this reality forces us to rethink our own place in the world, not as superior to other living beings but amid them. This perspective also allows us to discover the infinite richness of animal lives and the dazzling complexity of "beasts."

Enlightened by science, this series endeavors to open doors to these other worlds by providing a new understanding of living things and, therefore, a new understanding of ourselves.

Also in the Series

Jessica Serra, *The Beast Within: Humans as Animals* (2024)

What Do Bees Think About?

Mathieu Lihoreau

Translated by Alison Duncan

Johns Hopkins University Press
BALTIMORE

Printed in the United States of America on acid-free paper
9 8 7 6 5 4 3 2 1

This work was originally published in French as *À quoi pensent les abeilles*

Johns Hopkins University Press
2715 North Charles Street
Baltimore, Maryland 21218
www.press.jhu.edu

Cataloging-in-Publication Data is available from the Library of Congress.
A catalog record for this book is available from the British Library.

ISBN: 978-1-4214-4858-9 (paperback)
ISBN: 978-1-4214-4859-6 (ebook)

Special discounts are available for bulk purchases of this book.
For more information, please contact Special Sales at specialsales@jh.edu.

CONTENTS

SERIES EDITOR'S FOREWORD

Ethology is fascinating in many ways. Each species we study opens the door to a new world. But there are worlds so different from ours that they're beyond comprehension. Like navigators in search of unexplored continents, some scientists take up the challenge. Mathieu Lihoreau is one of them. Expanding our understanding of animal intelligence, he urges us to have greater respect for insects. His journey to the heart of a miniature world captivates as much as it challenges our certainties.

Some illustrious people before him had already begun to unravel the mysteries of these tiny creatures. Of all the insect societies that have been studied, none is more fascinating than that of bees. While in his *History of Animals*[1] Aristotle does not attribute *logos*[*] to animals, he does grant these social hymenopterans a remarkable intelligence that is superior to that of certain "blood animals."[2] According to him, the bee colony is a model for human political organization, which through careful study can shed light on the functioning of human societies. In those ancient times, when male dominance prevailed, scholars attributed the regency of the colony to a king bee and reasoned that the workers' incessant activity and division of labor had to be dictated by a male. Centuries later it was discovered that the winged monarch was in fact a queen, who didn't govern the workers, and that these workers had different functions during the course of their lives.

* According to Aristotle, *logos* is unique to human beings. Broadly speaking, it characterizes the distinct nature of human communication and community.

For Pliny the Elder, the divine bees "are patient of fatigue, toil at their labours, form themselves into political communities, hold councils together in private, elect chiefs in common," and, what is "most remarkable of all, have their own code of morals."[3] And therefore, "if the king should happen to be carried off by the pestilence, the swarm remains plunged in grief and listless inactivity; it collects no more food, and ceases to issue forth from its abode; the only thing that it does is to gather around the body, and to emit a melancholy humming noise."

Greco-Roman philosophy wasn't the only tradition to see a model of virtue in this noble insect. The bee is a key animal in the biblical bestiary and embodies the heart of work, devotion, and merit. The theologian Gregory of Nyssa used the bee as a metaphor to illustrate the work of the priests gathering the Scriptures: "In these words Proverbs counsels that one should not depart from any of the good teachings but, flying to the grassy meadow of the inspired words, should suck from each of them something that assists the acquisition of Wisdom and make oneself into a honeycomb, storing the fruit of this labor in one's heart as in some beehive, fashioning for the manifold teachings separate storage places in the memory, like the hollow cells in a honeycomb."[4] More than mere earthly food, the bee provides a spiritual provender. Seen as immaculate and a symbol of chastity, the queen, who is capable of reproducing without the help of a male, fascinates the defenders of virginity dogma.

In their remarkable book *L'abeille (et le) philosophe, étonnant voyage dans la ruche des sages* (The bee (and the) philosopher: A surprising journey into the wisdom of the colony),[5] the brothers Pierre-Henri and François Tavoillot explain that, like a mirror of humanity, the bee has been a symbol of different ideological issues. It was seen as libertarian and mutualist by the anarchist Pierre-Joseph Proudhon, as industrial by the economist Henri de Saint-

Simon, and as "the symbol of communism opposed to commercial liberalism" by the French president Adolphe Thiers.[6]

Rather than seeing the bee colony as a metaphor for humanity, Mathieu Lihoreau has stepped into the head of this royal insect. And the change in perspective is astounding. "Elements that were visible to us disappear . . . and others become more salient because they have an important biological significance." In the bees' realm, some of them smell the fragrance of flowers at their leisure, anchor this sensorial delight in their memory, mentally represent the aroma, and, back at the palace, pass on their delicious findings while dancing and waggling. They locate the precious nectar by situating themselves in relation to the sun and use other terrestrial visual cues, which they're capable of counting. And while the forager bees are intoxicated with subtle fragrances, others are involved in the manufacture of the perfectly designed nest. They don't do this in the automated way people have imagined but rather by adapting to the geometric constraints of a space, which requires a "mental-image template of the desired outcome." Then, when the time comes to build a new nest, some of them become daring. These Magellans of the air prospect even the smallest cavities, assessing every detail. Back at the colony, they disclose their findings. Then the debate begins in the land of honey. With varying degrees of intensity, each bee broadcasts the location it would choose, and "the final decision is based on the intrinsic superiority of the winning site, evaluated many times by several hundred individuals. A simple and effective model of democracy."

And what can be said about bees' extraordinary learning abilities? That they exceed the wildest hypotheses. Not only can bees "copy their peers," but they can also "improve upon observed behaviors." Mathieu Lihoreau even discusses the probable existence of consciousness in bees, since these hymenopterans have "the ability to recognize oneself as distinct from another entity, to plan, to recall specific events, and to take on the perspective of other individuals."

Through this richly documented book, Mathieu Lihoreau delivers a vibrant tribute to insects, showing that the world of miniature beings should envy nothing in the world of giants!

Jessica Serra

What Do Bees Think About?

PROLOGUE

What Is a Bee?

Have you ever observed an insect? I mean really observed one. By approaching it carefully. Then watching it for several seconds. Without disturbing it. Did you wonder what was going on in its head while you were watching it? Was it afraid? Did it carry on as usual? Did it even notice you?

Like us, insects have organs that enable them to see, smell, and touch. They also have a brain that takes up almost the entire volume of their head. But, unlike our brain, theirs is tiny. Reasonably, it makes us wonder what purpose it serves and how it works. Do insects have intelligence? Do they feel emotions? Are they conscious beings? Are they creative?

I've been studying these questions for about fifteen years. My relationship with these small invertebrates that have a head, thorax, abdomen, six legs, and (almost always) wings got off to a rough start, however.

I mistreated them at first. As a child, I lived in a house with a yard. Like many young children, I'd developed the habit of tasting everything I found on the ground. Often, I would catch ants. Other times, it would be ladybugs. This was much to the dismay of my parents, who fed me well. Then, as I grew up, I started playing war. I besieged anthills with my plastic soldiers and captured their ant leader. At that point, I had no idea that these insects have an immense devotion to their queens, protecting them at all costs and not hesitating to sting or bite any intruder trying to approach them. Later, I wanted to tame them. I would put ants in an empty jam jar and pierce holes in the lid to make sure they could breathe. Then I would add twigs to keep them from starving and firebugs to keep them company. Unfortunately for the latter, ants can be very aggressive, and they need animal protein to feed their larvae and reproduce.

When I grew up, I learned to love insects. It was at university that I discovered ethology, the science of animal behavior. During practicums, we would watch videos of animals with the task of writing down everything we observed in an attempt to understand their behavior. How many times did a young herring gull have to beg its mother for food to get a fish? How long would it take for Canada geese to decide to stop grazing on grass and look up to watch for predators? These videos were endless. But they sparked a vocation in me.

Little by little, I began to see the animals around me differently. I first tried to photograph birds. But as my camera's memory card filled up with failed shots, I realized that it was much easier to take great photos of insects. A bumblebee, for example, can remain perfectly still for several dozen seconds while foraging from a flower, its target two centimeters from its antennae. Crane flies (*Tipula*) are totally serene when sunbathing on a wall. Sure, I found insects to be photogenic, but not so photogenic that I wanted to spend my life studying them. Rather, I saw myself working with more charismatic animals, like great apes or whales. After all, the bigger the animal, the bigger their brain and the smarter they must be, right? But this shift didn't happen. And I'll try to explain why my path was inevitable.

I was introduced to research by my ethology professor. I didn't give much thought to the work I might be doing with him; I just needed an internship and a good grade. Alain Lenoir had spent his career studying chemical communication in ants. With him, I dissected hundreds of ants and analyzed their scents using techniques borrowed from physical chemistry.* I wasn't studying behavior. Or chimpanzees. But after several weeks of work, I was able to conclude that French ants had a different scent than Spanish ants, even though they were the same species. Since I hadn't read the scientific literature in detail (which one should always do before tackling a research question), I wasn't aware that I hadn't discovered anything novel. In most ant species, members of the same colony have the same scent, and this is partly determined by the many genes they have in common. Therefore, ants from geographically distant colonies (for example, on either side of the Pyrenees) are also genetically and olfactorily distant. Despite my stating the obvious, Professor Lenoir seemed satisfied with my work and encouraged me to continue in this direction. I did so with great enthusiasm, and that's how I came to be fascinated by the personality of cockroaches, how flies vote, how bumblebees navigate, how bees learn, and many other aspects of the private lives of insects that I relate in the chapters of this book.

Today I'm an ethologist. I study the intelligence of insects. Still not great apes or cetaceans, but it's a great pleasure to encounter a world that's paradoxically still largely unknown to us. We often marvel when looking up at the sky and wonder what might be up there and whether our existence has any meaning.† Lowering our eyes to the ground can also be a source of wonder and can teach us a lot about our world, as well as about ourselves.

* Gas chromatography is used to separate molecules in a compound by vaporization. It's often used with mass spectrometry to then identify the molecules by measuring their mass.

† When I wrote these lines, the media was abuzz reporting the French astronaut Thomas Pesquet's departure for his second stay on the International Space Station.

Until very recently, insects were considered to be devoid of any form of intelligence or, as the French philosopher and scientist René Descartes (1596–1650) proposed, to be animated by mere "reflex mechanisms."[1] We now know that all animals are endowed with certain forms of intelligence and that these evolve as the species evolves. Insects are no exception. They have an extremely rich cognitive repertoire that is sometimes puzzling to us. And this is all the more remarkable because their brains are ridiculously small compared to ours. This mental sophistication isn't the fruit of an isolated researcher's (my own) imagination but that of thousands of rigorous and independent scientific observations reported by hundreds of people for over a century.

In this book, I'll describe the different forms of intelligence identified in insects and discuss their limits. I'll mainly discuss species that have been observed very closely for several centuries and that I know well: bees.

But what is a bee? While the western honeybee (*Apis mellifera*), which has been domesticated since antiquity, often tends to steal the show, bees include nearly twenty thousand species and are the cousins of wasps and ants. This clade includes all species of honeybees (eight species that are also used for their wax, propolis, and crop pollination), bumblebees (two hundred and fifty species, some of which have also been domesticated for pollination), stingless bees (five hundred species that, as their name implies, do not sting), and thousands of species of solitary bees that are generally given less attention but whose importance is crucial to our environment.

This book is not an exhaustive list of everything we know about these bees. For that, I advise you to read the works of my fellow researchers[2] or practical guides for beekeepers.[3] This book is more a collection of scientific advances, anecdotes, and reflections on bees' intelligence and on the ethologist's profession in the 21st century. Many of these discoveries have been made with bumblebees. Although less well known than honeybees, bumblebees are social bees and very common in gardens. They're round and hairy and they're often viewed affectionately. They're easy to raise in shoeboxes and within just a few years can be used as model species for studying bee behavior in general,* similar to how laboratory rats are models for studying mammals. The team I lead in Toulouse, France, is dedicated to studying their intelligence.

To better understand these insects' psychology, we must try to project ourselves into their world as we know it, or rather as we imagine it. This is what ethologists try to do every day. Until proved otherwise, bees don't talk to us. They can't explain to us what they feel or what they understand about a situation (if they tell you one day, you'll probably have a Nobel Prize coming your way!). And while we're beginning to be able to translate human thoughts into words thanks to brain imaging and artificial intelligence, we're still far from being able to decode insects' thoughts. I'll therefore try to immerse you in the world of these creatures by relying on experiments that have allowed me and many other researchers to decipher different aspects of their behavior and to depict, little by little, their inner lives. An additional difficulty with beings so genetically distant from us is that their sensory system is also very different from our own. Their organs that are equivalent to our eyes, ears, nose, mouth, and fingers have capabilities that are foreign to us. The world of insects is therefore largely invisible to humans. The

* Bumblebees offer many advantages for research: they're domesticated, live in small colonies, aren't aggressive, and can be tested in the laboratory year-round. Three species are generally used: the buff-tailed bumblebee (*Bombus terrestris*) in Europe, the common eastern bumblebee (*Bombus impatiens*) in North America, and the Asian bumblebee (*Bombus ignitus*) in Asia. They're also widely used for pollination because, unlike honeybees, bumblebees are able to pollinate by sonication, also called "buzz pollination." They do this by vibrating their flight muscles to remove and collect pollen from flowers, pollinating them in the process. This form of pollination is necessary for the reproduction of many crops such as tomatoes and eggplants.

German biologist and philosopher Jakob von Uexküll (1864–1944) described this sensory space as a "soap bubble."[4] When we try to enter an animal's bubble, by using our imagination, our surroundings are transformed. Elements that were visible to us disappear as we enter the bubble, and others become more salient because they have an important biological significance for the animal. For example, I live very well without distinguishing the scent of a daisy from that of a dandelion, whereas this ability is vital for a bee. The bee's bubble allows it to smell the difference. On the other hand, I imagine that bees are unable to distinguish one of Mozart's symphonies from a Beatles hit. It's our human bubble that allows us to hear the difference. So, then, what is a bee's bubble like? What information does a bee get from its environment? And how does it use that information to make intelligent decisions? I'll answer these questions based on findings from scientific research. Everything you'll read is true—or at any rate, these experiments have a probability of being wrong less than 5% of the time.*

* By convention in biology, a result is considered significant (true) when a statistical test yields a probability of less than 5% that the result is false. If the probability is greater than or equal to 5%, the result is not significant.

1

A Poor Sense of Direction

This might be the case with you too, but I have no sense of direction. I always get lost wherever I go. Sometimes this annoys those around me. "Again? Are you doing this on purpose or what?" Even my five-year-old daughter criticizes me, which is embarrassing. But unfortunately, I'm not doing it on purpose. Let me explain.

I spent my childhood in Tours, France. During those years, I passed the same buildings, the same streets, and the same squares hundreds of times. All of these places are perfectly familiar to me. But I have no idea how to get to any of them or in which direction I should go when I leave someplace. Yet the city center was built according to an irrefutable logic. It's bounded by two rivers, the Loire and the Cher, with several perpendicular thoroughfares built between the waterways. This construction was an unfortunate consequence of the destruction the city suffered during the Second World War, but nevertheless it's useful for getting one's bearings. Historic monuments such as the cathedral, the castle, and the ruins of the basilica tower over most of the other buildings in the city, creating visual cues that make it possible to orient oneself from a distance. Despite an urban design that is conducive to efficient travel from place to place, I often get lost in my hometown, even when going to places regularly visited by tourists. What is perhaps the most annoying is that my poor ability to navigate isn't specific to Tours: it affects me everywhere I go, in every city I've ever lived in, whether large or small, a modern metropolis or an old town with winding streets. I can't do anything about it.

Fortunately, today's phones are equipped with GPS. I still have a hard time figuring out where I am, but I've stopped missing appointments and flights due to getting lost. All I have to do is scrupulously follow the route my phone shows me, only lifting my eyes from the screen to avoid obstacles in my path. So I make do with my problem and never really thought to wonder how it came about. That is, until the day I became interested in bees. Unlike me, bees have a reputation for being excellent navigators. Yet a bee's brain contains about one hundred thousand times fewer neurons than mine. And their life expectancy, during which they learn to navigate, is less than 0.1% of mine.[1] So how is it that these creatures get lost much less often than I do, without even the assistance of a smartphone?

Some insects have navigational abilities that far exceed our understanding. You probably already know about the huge migrations of monarch butterflies that cross North America to escape the Canadian winter for the warmth of Mexico. Year after year, these orange and black clouds invariably follow the same path even though they represent successive generations of butterflies. Or perhaps you've heard about the gigantic swarms of migratory locusts that fly over the African deserts in search of green grasslands.* Some dragonflies even have the ability to cross the Indian Ocean to reach Africa and the Middle East from India.

* Locusts and grasshoppers have haunted people's consciousness since the Bible identified them as one of the ten plagues of Egypt. Today climate change favors their gregarious behavior and invasive propensities.

Impressive indeed. But many insects navigate in a way that resembles our own. Among nest-occupying species, such as bees, ants, wasps, and many others, navigation involves learning to move from a fixed point (the nest) to nearby sites of interest, such as a food source or breeding area. It's similar to our routine of going from our home to our workplace or shopping areas. But for a bee that forages on a tree in bloom, getting the right orientation is a matter of life and death. Any failure to get from the nest to the tree, and then from the tree back to the nest, makes the bee vulnerable to an excessive expenditure of energy, a lack of food, the risk of being caught by a bird, or acute stress due to prolonged isolation from its colony. Studying these insects shows us that it's possible to develop extremely efficient navigation systems with very little information and few neurons to process that information. So how do they do it?

In a Bee's Head

To better understand the navigation challenges that a bee has to resolve, let's enter a world far different from our own. Almost all species of bees, whether small or large, wild or domesticated, solitary or social,* live in a nest.† Among these bees, the adults who raise the young (larvae) must learn to forage in order to feed them. Larvae in particular need a lot of the protein and lipids contained in pollen to grow into adulthood. The adults, which are mostly females, mainly need the sugars contained in nectar to acquire the

* Most bee species are "solitary," meaning that a fertilized female bee is responsible for raising her larvae alone. Different levels of sociality are identified when several females cooperate to raise larvae. In the most social species ("eusocial"), such as the western honeybee, labor is divided between the adults (queens and males), who produce new larvae, and the non-reproductive adult females (workers), who care for the colony.

† Only a handful of species do not build nests but lay eggs in the nests of other bee species, thus providing their larvae with shelter and warmth—an act of parasitism that earns them the name cuckoo bees.

energy necessary for domestic tasks. Reproductive females also need protein to develop their reproductive organs (ovaries) and produce eggs.

Therefore, foraging is a daily task for a bee. At first glance, it might seem straightforward. To our human eyes and nose, the flowers bees visit appear to be very colorful and fragrant. It's tempting to say that sight and smell should be a sufficient guide, but if we try to put ourselves in a bee's head for a few seconds (its "soap bubble," as von Uexküll said), we can see that foraging is much more complex than that.

This exercise requires a bit of imagination. On the one hand, you have to see the world in a very big way, through a pair of compound eyes. You fly through your environment by vigorously beating your wings, feeling it with your articulated antennae, and tasting it by using your legs and proboscis, all while moving at an average speed of 12 miles per hour. On the other hand, you might be afraid, as I am, of stinging yourself! Nonetheless, you'll quickly understand that getting your bearings and foraging under these conditions requires you to solve a certain number of mental exercises.

When a bee (you, in this case) leaves its nest for the first time, it must learn to locate itself so that it can return to this precise place, whether it found food on its route or not. After a few days or weeks spent in the nest, the bee undertakes training flights during its first outings, which resemble a three-dimensional dance. During these distinctive flights, the bee moves slowly in front of the nest entrance. It makes lateral movements, from right to left, wider and wider, then up and down, until it disappears over the horizon. From these controlled zigzags, the bee learns to locate the entrance to its nest. It's thought that the bee builds a visual memory of the scene, much like taking a picture of the position of the nest and any other prominent landmarks in the vicinity, such as branches, plants, or the skyline, which may be interrupted by mountains or houses. Once memorized, this constellation of visual cues can be used during its return flight to locate the nest at close range.

Several famous pioneers in ethology, such as the French entomologist Jean-Henri Fabre (1823–1915)* and the Dutch ethologist Niko Tinbergen (1907–1988),† have shown that it's relatively easy to trick these insects by changing the arrangement of visual cues at the nest entrance between the moment it leaves the nest and when it returns. Tinbergen's classic experiment involves surrounding the entrance to the nest with a wreath of pine cones and then moving the wreath a few centimeters away. This directs the insects' seeking behavior toward the pine cones and not toward the actual nest entrance. The experiment also works well with larger objects. I later learned this the hard way after staying a little too long watching bumblebee orientation flights while sitting next to their hive. Without my noticing, the bumblebees had gradually associated me with the entrance of their colony; so then for several days a group of forager bees had the habit of following me wherever I went, desperately looking for the entrance to their hive under my clothes.

Once these first flights are completed, the bee must locate flowers from which it can collect nectar or pollen. To do this, it uses its compound eyes. Each eye has an average of five thousand hexagonal photoreceptors (called "ommatidia") that collect information that is transmitted to the brain. But the bee does not see in mosaic;

* A naturalist and hero for many ethologists, including me, Fabre is known worldwide for his *Souvenirs entomologiques*, a series of texts on insects and arachnids, which combine both scientific observations and poetry. Today his hometown of Saint-Léons in France is the home of a tourist attraction and museum called Micropolis, the city of insects.

† A professor at Leiden University in the Netherlands and winner of the 1973 Nobel Prize in Physiology or Medicine, Tinbergen is considered one of the founders of ethology. He is particularly known for his four questions, which are categories of explanations for understanding animal behavior: mechanism, ontogeny, adaptation, and phylogeny.

it sees a single image. To continue your immersive exercise, you must now double your field of vision to see almost 300 degrees, instead of the 180-degree visual field of your human eyes, and you must see at a rate of 200 frames per second instead of 24. That is, you see neon light as a strobe. You're also severely nearsighted but have color vision. Like us, a bee sees color, but in a slightly different light spectrum, ranging from orange to ultraviolet. A bee is particularly sensitive to blue, green, and ultraviolet light, so most flowers appear to it differently than they do to us. Many flowers have ultraviolet patterns that are invisible to our eyes but that guide bees to the nectar-secreting glands at the base of the flowers (called "nectaries") that attract insects. For example, a forager bee will perceive a beautiful red poppy as gray with an ultraviolet spot at its base.

Bees are also equipped with two mobile antennae, which are sensitive to a wide variety of scents. The scents emitted by flowers allow bees to locate them. Each antenna has several thousand sensory receptors (called "sensilla") in which odorants are captured when the bee passes through a cloud of fragrance. Some high-abundance fragrant volatiles guide bees over long distances to flowering sites, while other volatiles with lower emissions are perceived only at short range when the bees arrive at the sites and must select the flowers to visit first. So, to take this exercise to its logical conclusion, you now have a relatively hairy, articulated "nose" above each eye. The "soap bubble" is taking shape.

Once a flower is located, the forager bee must judge whether it's a good source of food. For that, the bee is equipped with a gustatory system that enables it to taste nectar and pollen through contact with its legs (called "tarsi") and its proboscis. If the food is good, the forager bee learns to recognize all flowers of the same species by remembering the flower's visual aspect (color, shape, patterns) and olfactory aspect (scent). This is how the bee learns.

Other senses may come into play too. For example, Daniel Robert and his team at the University of Bristol in the United Kingdom have shown that bumblebees are able to use the natural electric fields around flowers to identify their species and the amount of nectar available.[2] These electrical signals, in the range of a few volts, come from the earth's electromagnetic field. Every plant in contact with the soil is surrounded by a negative field whose characteristics depend on its size and shape, like how the electric field surrounding you or me depends on our stature. Even though bumblebees do not have an organ dedicated to perceiving these electric fields, they are able to sense them. When a bumblebee passes near a flower, its own electric field interferes with the flower's, creating a current of static electricity that can tilt the hairs (called "scopae") on the surface of its body and provide information about the electrical signature of that flower. When a bumblebee is close to a flower, the difference in charge between the bee's positive field and the flower's negative field also allows the pollen grains to jump a few centimeters and stick to the bee's body. The bumblebee therefore acts as a pollen magnet, facilitating pollen dispersal. Even more surprisingly, a bumblebee's visit to a flower alters the plant's electric field for a few seconds, informing other pollinators that the flower has been recently visited and that its nectaries are empty.

To make its foraging journey worthwhile, a bee must visit several flowers on each trip. With each visit to a flower, a bee stores a few microliters of nectar in its honey stomach (also called the "crop"). The western honeybee's crop can contain about 60 microliters, or one third of its body weight. The bumblebee's usually can contain more than 120 microliters. Therefore, it's estimated that a bee must visit several hundred flowers per trip to fill its storage capacity. These flowers can sometimes be far apart from one another and be distributed irregularly across the landscape, as is the case, for example, in meadows. This means that bees must travel great distances to forage.

It's generally believed that a western honeybee forages within a radius of 3 kilometers, just under 2 miles. This distance is used as a reference by beekeepers when they want to move their hives to new environments, but it's probably an underestimate. Jürgen Tautz and his team at the University of Würzburg in Germany have shown, for example, that bees captured at a feeding site, transported in the dark, and then released 10 kilometers (approximately 6.2 miles) away are able to return to their original hive within a few hours.[3] Some euglossine bees (commonly known as orchid bees) can travel up to 20 kilometers (nearly 12.5 miles). In social species, such as honeybees and bumblebees, foragers may repeat these long routes dozens of times a day, for several days or even weeks. As feeding sites change over time—as flowers secrete more nectar or as trees produce new flowers—bees can learn to return to these sites, meaning that they develop just as many memories of new places as they do memories of known sites to forage. So when they discover a new site of interest, the forager bees take their usual zigzag learning flights, like they do to learn their nest location, but in this case they're memorizing the visual scene using the constellation of cues that characterizes the site where they discovered the new flowers.

As flight is a particularly energy-consuming mode of locomotion, it's believed that it's vital for a bee to learn how to reach these different feeding sites using optimized flight paths. Bernd Heinrich, a professor at the University of Vermont and an avid runner, has spent part of his career studying these energy concerns.[4] He especially had fun comparing a bumblebee flying to a human running. According to his calculations, bumblebees have one of the highest metabolic rates ever recorded in an animal. Heinrich thus estimated that, while running, a human burns the energy contained in a chocolate bar in about an hour, while a bumblebee of equivalent mass (a frightening image!) would burn the same amount of energy while flying in only thirty seconds. For a bee, foraging is therefore a considerable challenge. It requires the ability to learn and remember a large amount of information in record time about the location of the nest and the flowers to forage in order to minimize travel.

Everyone on the Dance Floor

The study of insect navigation is often believed to have begun in Germany nearly eighty years ago when Karl von Frisch (1886–1982), then a professor at the University of Munich, made the remarkable discovery that western honeybees use a form of symbolic communication to signal the location of food sources to other members of their colony.[5] This observation earned him the Nobel Prize in Physiology or Medicine in 1973, awarded jointly with Niko Tinbergen and Konrad Lorenz (1903–1989),* and marked the recognition of ethology as a scientific discipline in its own right.

Since ancient times we've known that western honeybee foragers returning to their colony with pollen or nectar perform what has been termed a "waggle dance"† in the darkness of the nest. This dance involves a waggle run straight along the surface of the nest followed by a turn to the right or left to return to the starting point. This movement is repeated several times in a row, resulting in a figure-eight pattern. It looks a bit like people dancing the Charleston or the Macarena, except that in this case there is only one dancer, and the dance floor is vertical. To decode the dance of the bees, von

* An Austrian biologist, pioneer of ethology, specialist of wild geese, and author of many significant discoveries, such as the concept of imprinting, Lorenz observed that chicks would hatch from the egg and follow the first moving object they saw. This object could be their mother or Lorenz himself, but either way, the chicks acted automatically, following anything that moved in front of them.

† Von Frisch used the term *schwänzeltanz* in German for this dance, which was translated as "waggle dance" in English.

Frisch used artificial flowers. In scientific studies, artificial flowers are commonly used instead of real flowers to control the variables of an experiment. While artificial flowers vary in design, they don't necessarily resemble actual flowers. Instead, imagine a reservoir filled with sugar water and a delivery mechanism from which bees can come and drink. With his artificial-flower system, von Frisch was able to train bees to forage on food sites whose location and quality he could control precisely, placing the artificial flowers wherever he wanted and filling them with whatever he wanted. Using glass hives, he was also able to observe the forager bees' waggle dance in great detail, as well as its variations after he moved the artificial flowers or changed their contents.

In the summer of 1945, von Frisch began to mark individual bees with a drop of paint on their thorax (roughly the equivalent of our trapezius muscle). This allowed him to discover that a bee that had been in contact with a "dancer" was able to go to the site indicated by the dance without ever having visited it before. Thus the dancers were communicating the location of a food source to the "follower" bees. After numerous observations and measurements of the different characteristics of this dance with a compass and a watch, von Frisch understood that the angle the bee waggled in relation to the vertical of the nest indicated the angle of the food source's location in relation to the position of the sun in the sky. If the bee was dancing vertically, it was indicating the direction of the sun. Von Frisch also understood that the waggle run's duration correlated to the distance from the nest to the food source. The longer the dancer waggled, the longer it would take to fly there. Finally, the number of times the dance was repeated provided information about the quality of the food source. The more nectar available, the more the bee would repeat its dance. Interestingly, if an artificial flower was placed close to the colony, say less than 100 meters (or less than 100 yards) away, the forager bee would do a "round dance,"* simply turning continuously to the right or left. This circular waltz indicated the presence of food outside the nest, but the information conveyed about its distance and direction was much less precise than in the waggle dance. During these dances (both the waggle dance and the round dance), the bees also transmitted information about the food source's scent, so that the follower bees would also be guided by their antennae to find the site in question.

It's now known that all honeybee species use the waggle dance to recruit their fellow bees for rapid and collective foraging of food sources. There are variations between species, however, which are called "dialects," just as there are variations in a human language according to geographical area. Even if we don't always understand other dialects, we often like hearing them and sometimes they even remind us of our childhood or our travels. But above all, dialects are direct proof that our cultures evolve. The same principle applies to bees. For example, to indicate a food source located 1 kilometer (about half a mile) away, a western honeybee's waggle run will be much longer than that of a dwarf or Asian honeybee. Depending on the species, the dances can occur in the dark or in daylight, vertically or horizontally, or with the position of the sun or the target itself as the point of reference. Other groups of social bees, such as bumblebees (belonging to the genus *Bombus*) and stingless bees (genus *Melipona*), also have forms of communication for recruiting other colony members to food sources. But these forms are much less complex, indicating only the presence of food outside the nest (as is the case of bumblebees) or requiring the follower bees to follow chemical traces deposited by the recruiter bees along the route

* The "round dance" is considered a distinctive version of the waggle dance and not an entirely different dance. The same behaviors and communication channels are involved but to different degrees.

(which is what stingless bees do). These variations most likely depend on ecological factors, such as the spatial distribution of resources, that are characteristic of the habitats in which the different species evolved. This means that the waggle dance would be particularly well suited to environments where resources are abundant but difficult to locate, such as the tropical forests where honeybees are thought to have originated.

A Vector of Misinformation

Bees learn to move between their nest and a food site and are able to memorize this route to communicate it to other members of the colony. This learning is done in several stages. Initially, bees use the sun and other visual cues, such as trees, the landscape, or buildings, to orient themselves. Then, as they gain experience, they learn movements that appear as "vector flight," which is movement in a straight line toward a particular target. This behavior is observed in classic displacement experiments. Once a bee has memorized a route from point A (the hive) to point B (an artificial flower), the experiment involves simply capturing the bee at point B and moving it to a point C that it does not know. To be absolutely meticulous, the bee should be transported in an opaque black box to ensure that it cannot see the route while being moved. When the bee is released at point C, it thinks it's at point B and automatically makes the B–A vector. It therefore ends up in a new location, point D, which is where point A would've been if the bee hadn't been moved.

This vector flight learning relies on cognitive processes that are common to bees and many other animals. After moving to a new city, it usually takes us a few days to get our bearings. We try several ways to get from home to work. Then one day we settle on a single route that we take every day from then on. It becomes a routine that we can do almost without thinking. Adopting a routine has the advantage of considerably reducing the cognitive load of navigation, which is the amount of information that needs to be processed in order to get around. But it can also be the source of errors. In my case, for example, that happens quite often. Every day, I cycle about 6 miles along the Canal du Midi between my home (point A) and my laboratory (point B). This path has become a routine. One evening, a colleague who lives halfway between my home and the laboratory (point C) was hosting a dinner party. So I left my workplace and followed the canal to my colleague's house. After dinner, I got back on my bike and pedaled the 3 miles or so from her house to my house, still following the canal. But when I arrived at the laboratory instead, I realized that my poor sense of direction had once again played tricks on me. When I should've turned right to go home (traveling C–A), I'd turned left (completing C–B). The direction was familiar and so was the distance. Only I hadn't used the correct vector.

Von Frisch's decoding of the waggle dance has provided ethologists with an extremely valuable tool for exploring the intricacies of bees' vector learning. You only have to read a forager's dance to know what it has learned. For example, several hypotheses were tested to understand how bees estimate the length of a vector. It had long been believed that bees do this by gauging their level of fatigue. But experiments that artificially increased the bees' effort by attaching weights of different sizes to their thoraxes were never able to demonstrate an effect. Today we know that to estimate the length of a vector, bees use the movement of images perceived on the retina of their compound eyes. This is called "optical flow." The more images a bee perceives per unit of time, the longer it estimates the distance it traveled to be.

The existence of this visual odometry mechanism was proven in a series of elegant experiments that continue to be models of simplicity and scientific precision. Harald Esch and John Burns at the

University of Notre Dame varied the optical flow perceived by bees by changing their flight altitude.[6] The researchers started from the observation that, given equidistant flights, a bee moving at ground level receives more visual information on its retina than a bee moving higher in the air because the ground is a significant source of optical flow. In fact, humans more easily perceive changes in the landscape while moving on the ground than in an airplane at an altitude of 10,000 feet. So in this experiment, bees that began their flight from a hive on top of a high building and were trained to fly to an artificial flower on top of another high building signaled a shorter travel distance in their dance than bees foraging between a hive and an artificial flower that were both set on the ground—even though the distance between the hive and the flower was exactly the same in both cases. Similarly, bees from a hive on the ground that were trained to visit an artificial flower suspended from a helium balloon about 100 meters (just over 100 yards) above ground level indicated a much shorter distance than when foraging on the same flower on the ground.

Mandyam Srinivasan and his team at the University of Queensland in Australia definitively put an end to the energy versus optical flow debate by training bees to feed on an artificial flower at the exit of a tunnel streaked with vertical black and white stripes.[7] It was as if the bees had to fly inside a giant three-dimensional barcode. By changing the width of the stripes on the tunnel walls, and thus the lateral optical flow perceived by the bees, Srinivasan and his colleagues observed that the bees signaled longer or shorter distances in their dances, even though in all cases the bees traveled exactly the same flight path between the hive and the flower.

Insects that move at ground level—and whose speed and optical flow variations are consequently lower—use other mechanisms to estimate distances. Ants, for example, count their steps. Rüdiger Wehner and his team at the University of Zurich in Switzerland demonstrated this by having desert ants* learn a route between their nest and a food source (these ants are crazy about cookies).[8] They were then captured at the food source and mounted on very small stilts made of hog bristles to increase the length of their strides (if you want to do the experiment at home, it also works well with toothbrush bristles). Once released in site unknown to them the ants on stilts made a vector walk toward the location of the nest, but they overestimated the distance compared to the ants without stilts and the ants whose legs had been partially amputated (which should not be done again!), thus proving these insects have an internal pedometer.

But let's continue to project ourselves a little more into the head of a bee. To know which direction to point a vector flight, bees use a "sun compass" by referring to the position of the sun in the sky. When the sun is not directly visible, such as on overcast days, they have access to other cues like the variations in the polarization of light in the sky. Sunlight, which is polarized in contact with the atmosphere, is perceived by bees thanks to three simple eyes, called "ocelli." These eyes are in addition to their compound eyes and are located on top of their head. Now you have a total of five eyes and can perceive electric fields.

As the sun moves across the sky, bees need to know the time of day in order to update the direction of their learned vector flights. Bees don't have a watch. But they do have an internal clock, which coordinates their physiological rhythms, such as the expression of

* There are several species of desert ants, but the best known belong to the genus *Cataglyphis*. These ants have excellent vision, and studying them has greatly contributed to the body of knowledge about the navigational prowess of insects. It all began with the discovery of a single ant that, after foraging more than 100 meters (more than 100 yards) away from its nest in a dry lake in the Sahara, was able to find its way back to its nest in a straight line with prey it had captured between its mandibles. To learn more, I recommend reading Rüdiger Wehner's excellent book *Desert Navigator: The Journey of an Ant* (Harvard University Press, 2020).

certain genes and hormone secretion. Von Frisch observed that it was possible to train western honeybees to forage at two different feeding sites and teach them that nectar was available at one site only in the morning and the other only in the afternoon.[9] This learning is only possible with a system that measures the passage of time. More recently, Randolf Menzel and his team at Freie Universität in Berlin, Germany, proved that disrupting a bee's internal clock could result in a misinterpretation of the sun compass and thus serious orientation errors.[10] The researchers first trained honeybees to forage on an artificial flower (this time an improved feeder with a ridged platform at the base to allow the bees to drink without drowning). They then captured the bees on the flower at 9:00 a.m. Half were anesthetized with isoflurane, which altered time perception by temporarily stopping the bees' internal clock. All the bees were then released at a new location unfamiliar to them at 3:00 p.m. Bees in the unanesthetized group moved in the direction the nest should have been by compensating for the movement of the sun, while the phase-shifted bees went in a different direction, as if they had not compensated for celestial movements during the six hours between their capture and release.

Estimating distances based on the movement of images on the eye and estimating direction based on the position of the sun in the sky are extremely important basic mechanisms that allow bees to locate themselves in relation to their nest at any time. This "path integration" ability acts as an onboard computer letting insects orient themselves in any situation. It's common to many animals, including humans. For instance, if you're made to walk a winding path blindfolded and then asked where you came from, you'll probably be able to point in the right direction.

The Traveling Salesman Problem

For a bee foraging in either an inflorescence, a flowerbed, or a tree, navigation is relatively simple. It just has to go from point A (the nest) to point B (where all the flowers are aggregated) and then back again to point A using a vector flight. On the other hand, if a bee is foraging on flowers that are distant from one another, like in a meadow, navigation becomes more complex. Then the bee must learn the location of several sites and an efficient route to connect them.

Finding the shortest path between several points is an optimization problem that mathematicians know as the "traveling salesman problem." It's called this because it considers the task of a salesperson who has to plan the shortest route to sell their products in several cities and then return to the starting point. In sales, time is money! It's therefore in their interest to pass through each city only once in order to avoid unnecessary detours. This problem is very popular in computer science because, despite its simplicity, there is no known algorithm that guarantees an exact solution. The only way to find the solution is with a brute-force approach, which involves calculating the length of all possible trips and comparing them. The number of trips is equal to the factorial of the number of cities to visit. A little reminder in case you haven't retained much from your high-school math classes: for n cities, we have $n! = 1 \times 2 \times \ldots \times (n - 1) \times n$ trips. This problem is therefore relatively easy to solve for a small number of cities but considerably increases in complexity as the number of cities increases. For example, if we consider that our traveler has to visit three cities (let's say Atlanta, Buffalo, and Chicago) and that the company is based in Atlanta, then there is a choice of six routes. Two of them have the shortest total distance, either in one direction (Atlanta to Buffalo to Chicago) or in the other (Atlanta to Chicago to Buffalo). Now let's say the salesperson needs to travel to ten cities. The number of possible routes increases to 3,628,800. For a bee that has to visit several dozen flowers, the problem becomes very complex, if not impossible, to solve without a computer server. So how do they do it without any assistance?

This question was first studied by Charles Darwin, the famous cofounder of the theory of natural selection.* Darwin spent the second half of his life experimenting with his theory at his home in Downe, south of London. In his notes from 1854 to 1861,[11] he explained that he spent long hours observing bumblebees, which the British then called "humble bees." Darwin suspected that bumblebees regularly revisited the same plants in the same sequence. He wondered whether these routes were consistent for a given bumblebee, whether they varied from one bee to another, and whether bees generally tried to optimize their routes, like business travelers. To avoid chasing bumblebees, whose speed can reach over 12 miles per hour, Darwin involved his children in his experiments, as he often would. He placed six of his children at strategic points in the yard with the instructions to shout when they saw a bumblebee pass in front of them. With this method, Darwin was able to reconstruct some flight paths, but he also quickly came to the conclusion that a rigorous analysis of these movements would require marking the insects individually, perhaps with a distinctive paint spot on the thorax. By doing that, it would be possible to understand whether the same bumblebee always returned to the same place or whether several bumblebees happened to fly similar routes. These questions remained unanswered for a long time.

* The major contribution of British scientist and explorer Alfred Wallace (1823–1913) is now recognized. Wallace was a contemporary of Darwin and independently reached similar conclusions about the evolution of species. In fact, this prompted Darwin to publish his work earlier than originally planned.

A few years ago, I had the chance to work with Lars Chittka, a professor at Queen Mary University of London and one of the leading bee specialists, as a member of his team. Together we set up a series of experiments using modern tools that would solve the problems Darwin had encountered one hundred and fifty years earlier.[12] Unlike Darwin, I had the advantage of working with specially bred bumblebees.* So I placed a colony of bumblebees whose history I knew in a small wooden hive equipped with a long transparent plastic tube that I'd pierced in different places to slide cardboard doors inside. This created an air lock that allowed me to control the flow of forager bees in and out of the colony and, therefore, to select the bees that would participate in my experiment. To identify the different bumblebees in the colony, I glued a distinctive colored, numbered tag on the thorax of each individual (these markers are commonly used by beekeepers to identify honeybee queens in large colonies). I wanted to make sure that the bumblebees stayed within sight and that their environment was as homogeneous as possible, so I set up this hive in a greenhouse on the roof of the university and covered the windows with white paint that obscured all the surrounding visual landmarks. Seen from above, London is a forest of bricks and steel that could have caused interference for the bees. Finally, to control the location and quality of the feeding sites, I devised my own artificial flowers. Unlike simpler feeders, my flowers were equipped with an electromagnet that allowed me to give the bumblebees access to a cup containing a few microliters of sugar water. That meant that when a bee landed on a flower, I could decide whether to fill it or empty it thanks to a control box with as many buttons and red and green LEDs as there were flowers to control. Once the device was installed, the inside of

* Bumblebee colonies have been used commercially since the 1980s for pollinating greenhouse crops.

my greenhouse looked more like the inside of Doctor Who's TARDIS* than a field of flowers. In a way, it would indeed allow me to go back in time to test Darwin's hypothesis.

I then presented the traveling salesman problem to the bumblebees. One by one, the bees had to visit six flowers, each only once, to fill their crop with nectar and return to the nest. As I watched the bumblebees carefully, I found that after only a few hours they all ended up using one of the two optimal routes out of the 720 total possible routes. What is particularly astonishing is that the bumblebees achieved this result without trying all the routes. Instead they seemed to develop their route by trial and error, converging on the most efficient solution, as a learning algorithm would. A few days after these results were published, the British newspaper the *Guardian* headlined a story somewhat embarrassingly with "Bees' tiny brains beat computers, study finds."[13] The claim is not entirely false—bumblebees do have small brains. But it's not entirely true either. Scientists often complain that our words are distorted by journalists looking for sensational headlines. In this particular case, we may have embellished the story a bit ourselves during the interview. Of course bumblebees cannot outperform computers. But that does not make their skills any less impressive. Personally I don't know if I would've found the route easily, even in a small greenhouse.

I continued this research with Juliet Osborne, then a researcher at Rothamsted Research in Harpenden, north of London. On an experimental farm there, a team of electronics engineers developed a one-of-a-kind radar system to record the flight routes of insects too small to attach with a GPS transmitter. This radar would allow us to carry out our experiment in real-world conditions. The radar

* The TARDIS (an acronym for Time and Relative Dimension in Space) is a hybrid time machine and spacecraft from the British science fiction series *Doctor Who*.

system stands several meters tall with two large satellite dishes. The first dish, which is the largest, sends an electromagnetic beam straight ahead. The second dish, located above the first, receives a modified version of this beam (a harmonic) that is reemitted by a transponder (an antenna only a few millimeters long). When this transponder is attached to an insect's back, you can record its position. The radar rotates to cover a one-kilometer radius (a little more than half a mile). The entire system is battery powered, and the data is collected on a server in a truck. At the time, the system took two engineers to operate it.

So there we were in a field, in the middle of England, with a team of students and engineers to test whether the bumblebee behavior observed in the laboratory would be the same in natural conditions. First we spent a lot of time removing all the dandelions and daisies in the field that might have distracted the bumblebees from the artificial flowers. Bumblebees, like all pollinators, have a sensory system that has been shaped over billions of generations to locate the most common flowers in their environment. In our region, they have an innate preference for purple. Once the field was free of natural flowers, we deployed our artificial flowers. This time, five flowers were arranged in a pentagon configuration and spaced 50 meters (about 55 yards) apart so that the bumblebees, whose visual acuity is not as good as ours, would have to explore the area to find the flowers and develop their routes. The flowers were equipped with cameras connected to computers, which were powered by electric generators, to verify the radar data. So there were two people in the truck to operate the radar, one person at the entrance to the bumblebee colony to select which bumblebee to test and to fit it with a transponder, and two additional people in the field to refill the flowers with sugar water once the bumblebee had visited them (and to remove the dandelions when they grew back).

After three weeks of bad weather and failed attempts, we finally managed to record our first bumblebee trip with the radar. As soon as the bumblebee left the colony with the transponder on its back, we started communicating via walkie-talkies, amid the noise of the radar and generators, to relay its position in the field and follow the experiment live. This lasted for almost an hour, until the bumblebee returned to the colony. During this period of time, the bee located three of the five flowers and clearly had not used an optimal route as it had done in the laboratory, often revisiting the same flowers that had been emptied on the first visit. We then let the bumblebee perform about twenty more trials, foraging for about six hours on the artificial flowers. Next, after negotiating with the team to stay a little later that Friday night (in the UK, Friday is often a slightly shorter workday that ends at the pub), we recorded the bumblebee traveling a new flight path. This new trip lasted only ten minutes.

After comparing the two routes tracked by the radar, it was evident that our bumblebee had learned to use the shortest route to visit the five flowers. But even more impressively, the bee was clearly optimizing its flight distances by moving in straight lines between the flowers as if it had a ruler and compass. That evening, heavy rain fell in Harpenden. We packed up the radar in a hurry and headed to the Pavilion, the closest open pub in the village, to celebrate. Our efforts had finally been rewarded. Above all, we could feel we were making a major breakthrough in our field—namely that insects could solve route optimization problems that seemed complex to humans. Although it's the essence of a researcher's job, knowing one is in the process of making an important discovery is actually rare. We're used to making small advances. But that day, we had taken a big leap forward.

By the end of the stay, we had compelling evidence that bumblebees are able to learn and optimize their foraging routes, and to modify them very quickly when a new food source is removed, added, or moved in their environment.[14] This implies spatial learning and memory capabilities previously not believed to exist in insects. Today we know that honeybees also use these optimized routes, as do many species of solitary bees, butterflies, hummingbirds, some bats, marsupials, and primates, all of which feed on plant nectar and thus face similar navigational problems. Of course, these species, which are sometimes far apart on the phylogenetic tree,* have learning processes and memory capacities that are likely very different from one another's. But the resulting behavior is comparable. This is called "convergent evolution."

Cognitive Map and Cognitive Load

As we saw, insects are capable of finding their way with great precision in complex environments with numerous feeding sites that are sometimes distant from one another and beset with obstacles in between them. Understandably, this makes us wonder how these miniature navigators visualize space.

This question has been hotly debated ever since von Frisch began his research almost a century ago. Field observations have shown that bees can use shortcuts to connect two familiar sites (two artificial flowers) that they learned to visit independently from the same point (the nest).[15] This behavior was used as an argument to suggest that they could construct a "cognitive map," meaning a mental representation of the connections, distances, and directions that link different points in space without having directly experienced them themselves. At any given time, it's as if bees could read

* This term refers to a graphical representation that shows kinship relationships and traces the evolutionary history of groups of living things. Each node in the tree represents the common ancestor of its descendants. Darwin was one of the first scientists to propose a history of species represented in the form of a tree.

a road map embedded in their brain. This concept of a cognitive map was introduced in the middle of the 20th century by Edward Tolman, a professor of psychology at the University of California, Berkeley, to describe rodents' ability to use shortcuts in mazes. In mammals, evidence of cognitive maps has been found in the form of specialized neurons in the brain, called "place cells," that activate when an animal moves to a given location.[16] This discovery earned John O'Keefe, Edvard Moser, and May-Britt Moser the Nobel Prize in Physiology or Medicine in 2014. Subsequently, several other cell types activated by spatial information were identified in rat, mouse, and bat brains, fueling the cognitive-map theory. In insects, however, these neural signatures have not yet been discovered. Colleagues from Randolf Menzel's team in Berlin pursued this line of inquiry by training honeybees to forage between their hive (point A) and an artificial flower (point B), and then having them passively retrace the route (A–B) while attached to a drone. Although the use of electrodes to record the bees' brain activity during the passive flight to the flower may have revealed some electrical signals, to date there is no convincing evidence of the equivalent of place cells in insects.

There is no consensus that insects build cognitive maps comparable to those of mammals. Keeping the brain alert to process information and develop memories is an extremely energy-intensive process for an organism. There is no room for the superficial in nature. In fact, natural selection tends to limit cerebral investment as much as possible when it's not vital for the animal. Therefore, it's expected that insects, with their extremely compact brains, will eliminate an unnecessary cognitive load. The study of bee movement via radar analysis shows that in many cases it's not necessary to use a complex cognitive construct to explain their navigational prowess. On the contrary, many observations suggest that bees use much simpler strategies. For example, the use of shortcuts can be explained by associations between visual scenes and vector flights. If a bee that is taken from point A and then released at point B discovers that it has not arrived at its target (point C), it will try to identify visual cues around its new location (point D). If it's able to, because the visual scene is familiar, then the bee seems to average the vector already completed (B–D) with the vector connecting the new scene to its target (D–C). This is the same as using a shortcut (B–C) without the need for a cognitive map. By extension, several visual scenes and their associated flight vectors can be integrated into a single spatial memory. This type of mental representation, based on a partial knowledge of the environment, would allow bees to learn vector successions and thus develop routes between several sites.

The difficulty of testing these hypotheses in natural settings lies in the impossibility of fully controlling a bee's visual experience acquired during its lifetime. When a bee is moved and released at a different location, even several miles away, it can't be ruled out that it has never visited that location before. Moreover, although the area for the experiment is often chosen to be as homogeneous as possible, it's never strictly the case. For example, soil is a mine of information for bees. Even above a lake or over a desert, there are useful visual cues for navigation. With my team from Toulouse, I carried out experiments in a dried rice field in Andalusia in southern Spain (the climate in Seville is much more favorable than it is north of London for this type of experiment).[17] These rice fields had a huge network of perpendicular lines drawn by paths and roads built by the farmers, constituting many potential visual landmarks for the insects. Using the same radar as before, we showed that bumblebees tend to follow these structures on the ground, following straight and angular flight paths. This behavior is observed in bees even if they make large detours to explore the environment, visiting flowers and returning to the nest, much like in the video game *Snake.**

More and more researchers are therefore trying to overcome these limits in controlling a bee's experience by studying insect behavior in virtual environments. In the laboratory, images can be projected onto the walls and floors, which makes it possible to manipulate the insects' visual experience before, during, and after learning takes place in a very precise way. Elisa Frasnelli, a researcher at the University of Lincoln in the United Kingdom, recently succeeded in teaching bumblebees to fly through a virtual obstacle and land on a virtual flower using optical-illusion effects in an enclosed area with a screen on the floor.[18] The classic capture, displacement, and release experiments carried out in natural settings then take the form of "tele-transportation" experiments in which the animals are moved virtually. The coupling of these new experiments with technology to record brain activity should help us to better understand the organization of spatial memories in insects' brains and to assess whether we have overestimated or underestimated their capacity to represent the world by associating these memories with the human concepts of cognitive maps and GPS.

* In this video game, which was very popular before the arrival of smartphones, the player controls a snake that grows and thus becomes an obstacle.

2

The Fragrance of Déjà Vu

Often science fiction is ahead of science. Sometimes the two collide. When I was a college student, my behavioral ecology* professor, Jean-Sébastien Pierre, understood that his audience of young biologists was much more captivated by stories about aliens than mathematics. Accordingly, he used Ridley Scott's masterpiece *Alien* to explain optimal foraging, which refers to animals' tendency to optimize each of their survival-related decisions by following the economic principles of cost and benefit. In this film series, Xenomorphs (the aliens) lay their eggs in human beings so that the larvae can develop by consuming their host from the inside. After incubating for only a few hours, larvae usually emerge from their victims' bellies, producing spectacular eruptions of hemoglobin on screen. Some very real insects have similar behaviors. In fact, one species of Australian wasp belonging to the genus *Dolichogenidea* was recently named *D. xenomorph*.† Fortunately for us, these parasitoid

* A discipline that emerged from ethology in the 1970s, behavioral ecology aims to understand the function (adaptive value) of animal behaviors in their natural environment by borrowing concepts from ecology and evolution.

† Australian entomologists have a habit of borrowing names from pop culture. In 2019, Andrew Austin and his team at the University of Adelaide named two newly discovered parasitoid wasps after the Zygons, an alien race in the *Doctor*

wasps are much less scary than their alien namesakes, and they only attack other insects. What Professor Pierre wanted us to understand with this comparison was that, for these wasps, the decision to lay eggs in a host is too important to be made at random. It depends on many factors, such as the size of the host, the condition of its body, and the possible presence of other eggs. All these factors will affect the development and survival of the larvae. Although it may appear that the aliens pounce on the first human that comes along, in principle the same optimization rules apply to them as well.

A few years later, I discovered *Mimic*, another science-fiction film series that unfortunately didn't enjoy the same success as *Alien*. Nevertheless, the parallel with insect intelligence research is equally interesting. In *Mimic*, an epidemic transmitted by cockroaches ravages Manhattan, to the point that it's claiming thousands of children's lives. Since the infected insects have developed a resistance to pesticides, attempting to solve the problem with chemicals won't work. An entomologist in New York has the idea, however, to create genetically modified insects from termites and praying mantises that could mingle with the infected cockroaches to eradicate them. So far the storyline makes sense since termites and mantises are indeed related to cockroaches within the order Dictyoptera. In the film, these mutant insects, which the scriptwriters call the Judas Breed, release an enzyme that accelerates the infected cockroaches' metabolism, thus causing them to die prematurely. This is a textbook case of biological control: the genetically modified insects are sterile, so they can't reproduce in the wild, thereby reducing the risk of a biological invasion (bravo to the scriptwriters!). But, in spite of these precautions, the Judas Breed, which was initially intended to

---

Who series (*Choeras zygon*), and after chocolate-and-cream sandwich cookies (*Sathon oreo*).

live for only a single generation, ends up multiplying after they mutate into a humanoid form (don't ask me why).

During the final battle between humans and mutant insects, a small group of humans trying to save New York has the good idea to cover their bodies with secretions collected from the Judas Breed corpses. They understand that the Judas Breed, like many real insects, mainly uses scents to communicate. This ruse makes the now fragrant and very slimy humans invisible to the mutants since they smell just like them. The humans, because of their chemical camouflage, are able to destroy the mutants. In this film, Guillermo del Toro attributed a form of social intelligence to Dictyoptera, albeit fictitious mutants, allowing them to recognize each other and to organize themselves to colonize Earth. A few months after seeing *Mimic*, I demonstrated that this wasn't just science fiction. But I'll come back to that.

I should start by explaining what social intelligence is and why it's so important for living in a community. As social animals ourselves, we call upon it every day, every time we meet another person. We first recognize this person as a human. Then we identify their physical characteristics, such as their height, age, ethnicity, and so forth. Our ability to recognize faces allows us to distinguish between individuals, especially in a group. If someone is familiar to us, we can recall our shared connections and possible common experiences. We can then choose to spend more time with some people than with others. We can choose to tell them our innermost secrets and feelings, or we can choose to keep that information from them. Our social intelligence also allows us to learn from each other. Human civilizations are based on cultures built upon several thousand years of social transmission, which was first oral and then written. For insects, social transmission works in practically the same way. Let me explain.

The Scent of Hydrocarbons

BEE 1: "Bzz. Bzz bzz . . ."

BEE 2: "Bz."

BEE 1: "Bzzzbzzzzbz bzz."

BEE 2: "LOL."

What might two bees say to each other when they meet? Potentially many things. Bees make sounds. They can see, touch, feel, and perceive the electromagnetic fields around them. But, unfortunately, we haven't found a Rosetta stone* yet that would allow us to translate these behaviors into language or intentions that we could understand. This is precisely the challenge of the ethologist's work.

When two bees meet, it's essential for them to learn who they're dealing with. If they're in the same colony, they may have an interest in sharing food or information. If they come from different colonies, however, these bees have to decide whether fighting is worthwhile. Edward Wilson (1929–2021), a professor at Harvard University and the father of sociobiology,† considered this ability to self-identify to be a "social glue" that holds together the members of a society and shapes its organization.[1] For insects that live in complex societies, such as social bees, social wasps, ants, and termites, social recognition most often identifies the individual's role in the colony and membership in the nest. As a matter of fact, insect colonies can contain many tens of thousands, or even millions, of in-

* The Rosetta stone is a fragment of a stone tablet engraved with three versions of the same text in hieroglyphs, Egyptian demotic script, and ancient Greek. The discovery of this stone in 1799 made it possible in the 19th century to decipher the ancient Egyptian hieroglyphic system.

† Sociobiology is a discipline that studies the biological basis of social behavior. Much of the foundational research for this theory was obtained through the study of social insects and has subsequently been observed throughout the animal kingdom. Whether sociobiological concepts are applicable to humans is much more controversial, however.

dividuals. The members of the colony aren't all the same, and groups called "castes" can be identified. Insects belonging to the same caste all perform a particular role.

In western honeybees, the division of labor is taken to the extreme.* The colony includes a single reproductive female, called the "queen." She lives for several years and is the one who lays the eggs. There are also males, called "drones," whose only role is to mate, and sterile females, called "workers," whose life expectancy varies from a few weeks to a few months. It's these workers—the bees that we see foraging in our gardens—who make up the majority of the colony. These individuals undergo physiological and neurological changes throughout their lives that result in a progression of behavioral changes with age, called "age polyethism." In western honeybees, this type of polyethism is well regulated. The older a bee gets, the riskier the tasks it performs far from its colony. A young worker (in her first 3–4 days) cleans the honeycomb that will house the larvae and then the honey and pollen reserves. Then the bee feeds her queen and the larvae (days 5–10), produces the wax seal that covers the honeycomb cells (days 11–16), stations herself at the nest's entrance to ventilate it if it's too hot (days 12–22), and finally becomes a forager (days 23–30) until she dies after about a month of loyal service to the colony.

The western honeybee has fifteen glands that allow it to produce compounds containing useful information for communication (called "pheromones"). Since the early 2000s, much progress has been made in decoding these signals. For example, we know that in

* Many insect societies are characterized by a reproductive division of labor. This is one of the three criteria that define the most integrated level of sociality in living things (eusociality). The other two criteria are cooperative care of the young and overlapping adult generations in the group. Eusociality, which requires these three criteria, is a social system rarely found in the animal kingdom. Termites, ants, some bees, and some wasps are eusocial. A few species of aphids, shrimp, and mole rats also meet these criteria.

order to be recognized among the crowds of anonymous worker bees, the queen produces a royal pheromone secreted by glands at the base of her mandibles (the jaw's claws). The royal pheromone attracts the workers, who feed and clean her. As they do so, the workers cover themselves with the queen's scent, which they distribute to the other members of the colony through direct contact. This royal pheromone is essential for the proper functioning of the society. Among other things, it inhibits workers from constructing new queen cells, so there will be no other competing queens, and it reduces the development of the workers' genitalia (their ovaries), which prevents them from laying eggs themselves. Alison Mercer and her team at the University of Otago in New Zealand have shown that this pheromone also affects bees' capacity to learn.[2] When worker bees were exposed to the royal pheromone for a few minutes, the pheromone blocked their aversive learning, meaning the bees were unable to learn to associate an odor with a negative event, such as an electric shock, while they could detect the queen's scent nearby. On the other hand, these same workers were able to associate an odor with a positive event, such as a sugar-water reward. This is most likely a protective mechanism to prevent the workers from associating the presence of the queen with a negative event and suddenly stopping to care for her. The royal pheromone, essential for maintaining colony cohesion, has been documented in many other social insects, such as wasps and ants.[3]

Other chemical signals that cover the surface of the bee's body (called the "cuticle") allow the workers to identify their membership in the colony. These molecules are lipids called "cuticular hydrocarbons." Like petroleum and natural gas, they are organic compounds made up exclusively of carbon and hydrogen atoms.* The

* For biochemistry buffs: this hydrocarbon layer is primarily composed of linear alkanes, methyl-branched alkanes, and alkenes, with chain lengths generally ranging from twenty to forty carbon atoms.

primary function of these lipids is to serve as a protective layer of fat that allows the insects to avoid getting wet but also to avoid drying out. Smaller animals are particularly prone to desiccation because their body surface area, and therefore the area where water can evaporate, is proportionally larger in relation to their size. The cuticular hydrocarbons also provide physical protection against attacks by pathogenic bacteria and parasites that can't easily penetrate the insect's body. Finally, because these molecules are on the surface of the body, they are a mine of information that insects use to communicate. The diversity of compounds present on the cuticle of an insect combined with their relative proportion form a chemical signature, which provides information about the insect's identity. Like ancient lecanomancy, which divined the future in oil patterns,[*] insects are indeed capable of reading the identity of their fellow creatures in hydrocarbons.

In bee and ant colonies, surface hydrocarbons are exchanged continuously when individuals are in physical contact with their nest, when the antennae touch, and during the transfer of food (called "trophallaxis").[†] Through these chemical transfers, sisters, half sisters, and lesser-related workers in the same colony all end up carrying the same scent. This mixture, resulting from the homogenization of all colony members' individual scents, is unique and allows colonial recognition. This means that two colonies of the same species usually have the same mixture of cuticular hydrocarbons but in different relative proportions, allowing them to be distinguished from one another. As with cocktails, it's all about the mixture! If a foreign ant is introduced into an anthill, it will be immediately excluded or executed by the workers because its scent is

* This divination technique uses a dish of water or oil. It involves dropping in precious stones and predicting the future based on the ripples or reflections of light produced.

† This involves regurgitation of predigested food from the mouth apparatus of one worker into that of another.

slightly different. If the ant is first steeped in the host colony's odor, however, the ant will be accepted.

In western honeybees, discrimination happens at the entrance of the colony by specialized workers called "guard bees." They are the ones who protect the nest from dangerous intrusions, such as raids from neighboring colonies to steal honey reserves during periods of scarcity. The guards recognize intruders by comparing the scent of the individual who enters the colony with their own scent. The more similar the two scents are, the more likely the guards are to let the individual enter. Patrizia d'Ettore and her team at the University of Paris 13 have worked extensively on the subject of colonial recognition. They demonstrated that the odor of nest materials is a key component of bees' chemical identity.[4] In particular, the wax secreted by the worker bees for building the honeycomb cells consists mainly of fatty acids and hydrocarbons, which the bees can use to cover themselves as they wander around the nest. To demonstrate this, d'Ettore swapped the wax frames of different hives. She observed that workers from these different colonies had a greater tendency to accept each other than workers from colonies whose frames hadn't been swapped. Colonial odors in the paired hives became homogenized through bee contact with the waxes.

This chemical recognition system, although simple and efficient, isn't perfect because it depends on the level of similarity between the scent of the intruders and that of the colony. Depending on the varying thresholds of tolerance among guards, this level may be deemed acceptable or unacceptable. For example, in a western honeybee apiary where several colonies are placed side by side, up to 30% of the workers in one hive may come from another hive. If the hives are arranged in a line, this percentage is greater in the hive occupying the central position than in the hives at either end.[5] This drift of individuals is probably accidental, due to the bees' navigation or hive-recognition errors. It's permitted because the scents of bees in neighboring hives are close enough to be accepted. This acceptance is an attribute that beekeepers know well and make use of when transferring bees from one hive to another, either because there are too many bees (to avoid swarming)* or not enough (to keep the colony alive). Every apiarist has their own method. For example, my beekeeping instructor sprays diluted pastis (the French anise-flavored liqueur) into his hives to mask their colonial odors. There is no word on whether this works with other spirits, or whether the bees come back for another taste. But it's undeniable that this perfuming increases the acceptance rate of foreign workers in the recipient hives.

This phenomenon of acceptance is most pronounced when the bees are young, shortly after their metamorphosis into adults.† At this stage, they have no (or few) chemical signatures and therefore do not display their colonial origin. These bees then gradually acquire the scent of their adopted colony through contact with the nest and other bees. Some parasites take advantage of this security flaw in the colonial recognition system to enter bee colonies and sometimes establish themselves permanently by acquiring the colony odor. This is the case of cuckoo bees, whose females lay eggs in the nests of other bee species. Some of the female cuckoo bees have a reduced scent, which makes them chemically neutral or invisible to their hosts and greatly facilitates their introduction into the nest being parasitized.

Although colonial recognition is the norm in social insects, some have lost this ability, allowing neighboring colonies of the

* In western honeybees, whose queens can live for several years, swarming indicates the colony is dividing. The old queen departs the colony, taking half the workers with her, to a new nesting site, leaving the hive, nest, and food supply to the new queen with the rest of the workers. It's reproduction by fission.

† Bees are "holometabolous" insects, meaning they go through a complete metamorphosis between the larval and adult stages. This metamorphosis progresses through four stages: egg, larva, pupa, and imago (adult).

same species to join theirs with no distinction made. In these particular populations, workers, queens, larvae, and food are exchanged within huge networks of colonies constituting a single "supercolony." Notably, this exists in several species of ants. For a long time, the Japanese red ant (*ezo-akayama-ari* in Japanese) had the largest known insect supercolony. Located on the island of Hokkaido, the supercolony is estimated to be nearly 3 square kilometers (a little under 2 square miles), comprising more than two hundred million workers and one million queens distributed in forty-five thousand interconnected nests. But today, the record is held by the Argentine ant (*la hormiga argentina* in Spanish). You've most likely already encountered one of these ants. In its area of origin in South America, this ant forms colonies with a single queen, which is called "monogyny." It has a strict colonial recognition system, which usually results in killing any intruding ant. Since the beginning of the 20th century, however, the Argentine ant has been inadvertently imported into many countries, where it has successfully established itself. Laurent Keller and his team at the University of Lausanne in Switzerland identified a supercolony of these ants along the Mediterranean coast.[6] The supercolony extends over 6,000 kilometers (approximately 3,700 miles), from Portugal to Italy, passing through Toulouse (it's in my laboratory, a detail I'll elaborate on later*), and contains billions of workers and millions of nests and queens. These invasive populations are therefore polygynous, because their nests have multiple queens. Experiments mixing individuals from the supercolonies of Argentine ants in North America, Europe, and Japan have shown the inability of these ants to discriminate between members of their own locale and members of the other locales, suggesting that they belong to a single transcontinental "mega-colony." Of course, it's not technically the same colony because there is no

* See chapter 4, The Superorganism.

natural exchange of individuals across the oceans. But what this demonstrates is that all these invasive populations have lost their capacity for colonial recognition. This is perhaps the key to these successful invasions.

The Misunderstood Cockroach

It's thought that bees and ants aren't able to recognize their true degree of kinship, but only their membership in a colony, in order to avoid fratricidal conflicts that would destroy group cohesion. To try to understand this, you'll have to bear with me. A colony of insects should be thought of as a family. And as in all families, there are conflicts. Within this family of social insects, everyone helps each other to ensure the survival and reproduction of the youngest (the eggs and larvae). This form of altruism is a basic principle in the animal world. William Hamilton (1936–2000), a professor at Oxford in the United Kingdom and a leading evolutionary biologist,* named this "kin selection."[7] When the family is small, for example if the colony was founded by a single queen, all its members are sisters; or if the queen has mated with several males (a western honeybee queen can mate with about fifteen males), then the colony's members are half sisters. Recognizing your kin is therefore like recognizing the members of your colony. But when the family is large, for example if the colony was founded by several queens, the genetic diversity among the colony members increases considerably. Then there are sisters, half sisters, cousins, and many nonrelated workers. In this case, recognizing your relatives makes it possible to distinguish between these different classes of individuals

* Considered by many to be the Darwin of the 20th century, William Hamilton theorized many concepts in evolutionary biology that proved to be valid and fundamentally significant much later. The kin selection theory is the basis of sociobiology.

in the colony. But in these large families, if everyone tries to favor their closest relatives, the observed altruistic behaviors can go against the collective interest. Are you following me?

This phenomenon is even more pronounced in hymenopterans, the large order of insects that includes approximately 115,000 known species of bees, ants, and wasps. These insects have the genetic characteristic of being "haplodiploid," meaning that males have only one set of chromosomes while females have two sets (like you and me). In social species, only queens lay fertilized eggs, which develop into females—that is, the future queens and workers. But females are also able to lay unfertilized eggs, which are future males. This is called "parthenogenesis," a form of natural cloning. In insect colonies with a single genetic lineage (small families with one queen), workers may occasionally lay male eggs. But these are immediately destroyed by the worker "police" who either eat the eggs or remove them from the nest. Besides the fact that insect eggs are an excellent source of protein, the reason for this worker policing lies in haplodiploidy. On average, the workers are less related to the sons of other workers (their nephews) than to the sons of the queen (their brothers), so the first are eaten or removed before they hatch, while the second are spared. Are you still following me?

If we now consider colonies with multiple genetic lineages (large families with multiple queens), workers with the ability to recognize their true kinship could begin to destroy the male and female eggs of the queens least related to them. Such a repressive system would be counterproductive at the collective level and would no longer allow the colony to function properly because the workers would spend their time laying and killing eggs instead of building, feeding, and cleaning. So it's probably best for all to limit these insects' social recognition abilities to the colony level.

But not all insect societies are based on a system of division of labor, as is the case with bees and ants. Truth be told, most insects live in much less organized groups. Therefore, in these simpler societies in which individuals don't depend on each other for food and reproduction, kin recognition shouldn't pose a risk of harmful genetic conflict. I tested this hypothesis during my PhD program with Colette Rivault, then a researcher at the University of Rennes in France and one of the few (very few) specialists in cockroach behavior.

Before I joined her lab, I didn't know much about cockroaches, or roaches, call them what you will. But I did know about their bad reputation. And so, like many students before me, I wasn't particularly excited at the prospect of spending several years with cockroaches, especially as they generally elicit more of a desire to squash them than to study them. But my feelings quickly changed. Cockroaches' reputation comes from the fact that a dozen species live unwelcome in our homes. They're a bit like obligate parasites that entirely depend on humans. They rely on us to such an extent that it's uncertain whether these species could survive in the wild. Although it may be both very unpleasant to share your kitchen with cockroaches and extremely difficult to get rid of them, these insects are the targets of a number of urban myths that I must dispel before going any further. Cockroaches do not transmit diseases to humans. They do not release all their eggs when they're crushed. They do not nibble on our hair while we sleep at night. And despite their ability to survive in extreme conditions, they're not resistant to nuclear explosions. There are nearly forty thousand species of cockroaches[8] with varied morphology and habits. Some are colorful, and others sing a bit like cicadas. The majority live in tropical forests. Many are gregarious species in the sense that they live in fairly stable groups at different times of their lives. Some have developed close bonds with their offspring, even going so far as to feed their larvae with nutritional secretions, almost like nursing a baby. A handful of species have even acquired the ability to feed on wood, evolving into the huge group of social insects known as termites.* All these reasons make it a group of insects that can't be ignored when one is interested in the evolution of sociality. But not to worry—in this chapter I'm not going to ask you to project yourself into the mind of a cockroach. Instead I'll just explain how they communicate.

To test whether cockroaches are able to recognize their degree of kinship when they meet, and not just their group membership like bees, I took a close look at their sex lives. According to models of optimal foraging (the same ones for aliens and parasitoid wasps), animals choose their mates in a way that optimizes the quality of their offspring. Incestuous mating, especially between siblings, is generally avoided because the resulting offspring wouldn't be viable due to genetic incompatibilities. In cases of inbreeding in which young develop, they don't survive long or are unable to reproduce. After reading this, I decided to test whether cockroaches would avoid mating with family members when given the choice.[9]

On Colette's advice, my first step was to collect cockroaches out in the field and breed them. While the other students in my class were off to conduct their experiments on birds in Antarctica or horses in Mongolia, I went to trap cockroaches in infested buildings. To introduce myself there, I explained to the residents that I was testing new pest-control techniques for an extermination company (curiously, in addition to getting rid of rats, exterminators are often also specialized in "de-cockroaching"). This is certainly not the part of my research I am most proud of. But it was really only a half lie because, since so little is known about cockroach behavior, any new study, even on kin recognition, could only provide more useful knowledge to improve extermination techniques. At least in the long term.

* Phylogenetic studies based on molecular analyses show that termites (order Isoptera) are a branch of cockroaches that have developed social characteristics comparable in complexity to those of social bees, ants, and wasps.

On the rare occasions that someone let me into their home, I was able to set traps of my own making, made of bread and beer. Cockroaches love dark beer. Still following Colette's recommendations, I made a "cockroach catcher" out of a vacuum cleaner with a sock placed inside its tube to maximize its trapping ability. The vacuum cleaner collected the live cockroaches. The sock was essential to prevent the cockroaches from coming back out of the vacuum (use a longer stocking if you want to catch a lot of them at once). After several nights of trapping cockroaches in my *Ghostbusters*-like outfit, I finally succeeded in establishing a breeding setup for German cockroaches (the most common type in much of the world), keeping careful track of the identity of each individual, its place of capture, and its mating history. To do this, I had to mark the cockroaches one by one with paint. Each female and each male had its own color. Their offspring were marked with both colors. After only a few generations, the insects' entire bodies were covered with paint spots, which detracted from their original appearance but didn't prevent them from courting each other.

One day, I finally introduced females and males, whose parentage I knew, in transparent boxes; and then I observed them day and night. I had mentioned the idea for this experiment to Colette a few weeks earlier, but she hadn't seemed convinced of the need to invest this much effort in order to answer our question about incest. In hindsight, she was right: I could've thought a little longer and found a less onerous protocol. But I was eager to get started. So I didn't tell her everything. At least not at the beginning.

I started this experiment one Sunday morning when the laboratory was empty. By the middle of Sunday night going into Monday, I had realized three things. First, I learned how tiring it can be to watch cockroaches for twenty-four hours under a red light.*

* Cockroaches mate at night. Like bees, they have color vision, but their visible spectrum is shifted compared to ours so that red appears gray to them. This

Second, the experiment was already offering results. The cockroaches preferred mating with an unrelated partner to mating with a sibling. Finally, the rate of mating was low, so I'd probably have to monitor my boxes for several more weeks to gather enough data. After a little negotiating on Monday morning, we decided that Colette would cover days and I would cover nights.

During these long hours of observation, I took advantage of the breaks between cockroach matings to educate myself a little, reading Franz Kafka's *The Metamorphosis* and watching David Cronenberg's *The Fly*. My story ended less dramatically than these, however. By observing hundreds of cockroaches copulating in different types of boxes and labyrinths, we were able to show that these insects were able to recognize their precise degree of kinship to their potential partners, whether they were siblings, half siblings, cousins, from the same extended family, or from more distant populations collected in Rennes or Tours in France, or in Russia.[10] Moreover, this kin recognition gives structure to cockroaches' sociality. Males and females prefer to mate with strangers at night. But when it comes to choosing a safe site to remain away from predators during the day, they primarily gather in families. By huddling together, cockroaches keep warm and use less energy. They also avoid drying out by reducing the surface area of their body that is in contact with the air. They exchange beneficial microbes for digestion and information about available food sources in the vicinity, and they decrease their individual risk of being caught by a human or dog. It's therefore thought that cockroaches forming these kin groups is an altruistic act because they share the benefits of group living primarily with those whose genes they share in common. As with bees and ants (and the Judas Breed in *Mimic*), this social recognition is achieved through cuticular hydrocarbons. With cockroaches, how-

makes it possible to observe insects in the dark with a red light without disturbing them since red is invisible to them.

ever, chemical signatures do not homogenize within groups, so each cockroach retains its own scent, its individual chemical signature.

Liz Tibbetts and her team at the University of Michigan have shown that some social insects are capable of even more precise levels of recognition than those found in cockroaches. This is the case with *Polistes* wasps, which live in small societies in paper nests in which the members show complex and stable interactions over time. These social interactions require the ability to distinguish between individuals in the colony. In the species *Polistes fuscatus*, common in North America, colonies are often founded by several females that are not necessarily related. These female wasps fight one another to establish a dominance hierarchy that determines who will reproduce (becoming the queen) and who will work to develop the colony (becoming the workers). In just a few days, the amount and intensity of the aggression among the wasps decreases as the hierarchy in the nest stabilizes. The relationships become clearer.

While handling these wasps, Tibbetts noticed that they had distinctive facial patterns or at least a wide variability in head color and shape, even within a single nest. Some wasps were entirely brown, others predominantly yellow. But most of them fell between these two extremes. Tibbetts hypothesized that these patterns were the equivalent of our facial features. To test her theory, she had the brilliant idea to modify their markings with black paint. She then observed that she could alter the rate of aggression between wasps.[11] The altered wasps received more aggression than the unaltered wasps. The aggression diminished as the new markings became familiar to all the wasps in the colony. These insects' specialization in facial recognition allows them to recognize wasp markings faster than any other image, whether images of abstract geometric shapes or of caterpillars, their favorite prey. In contrast, other species of *Polistes* wasps, which are close to *Polistes fuscatus* on the phylogenetic tree but are solitary, are unable to learn the faces of other wasps. This demonstrates that individual recognition is an adaptation to group living, under certain conditions, and not an ability shared by all insects.

Golf, Robbers, and Peep Shows

Living in a group, in whatever form, allows us to learn from each other. We acquire new knowledge and skills from our fellow human beings, which we in turn pass on to others. It's the accumulation of this knowledge that is the basis of our cultures. Science is built on this same principle.* If Hans Lippershey (1570–1619) hadn't documented his invention of the telescope, Galileo (1564–1642) probably wouldn't have discovered that the earth revolves around the sun, and the Perseverance rover would never have landed on Mars to search for evidence of earlier life. Humbly I'd add, if several generations of researchers hadn't documented their work on chemical communication in bees and ants, I'd never have discovered kin recognition in cockroaches. I probably would've been too busy trying to figure out what their antennae are for in the first place.

In the 1950s, several researchers observed that the capacities for social learning and cultural transmission were not limited to humans.[12] Populations of blue tits in England have learned how to pierce the aluminum lids of milk bottles to drink the contents. Groups of macaques in Japan routinely wash sweet potatoes in streams before eating them. Chimpanzees in Tanzania shake hands to greet each other. Until very recently, it wasn't suspected that this type of cultural transmission could also be observed in insects. It is the case, however. And bees exploit this ability.

* In one of his letters, the British physicist Isaac Newton (1642–1727) wrote, "If I have seen further it is by standing on the shoulders of Giants." This statement is now commonly used to describe how science builds cumulatively on previous discoveries.

Lars Chittka, my colleague at Queen Mary University of London, has contributed a great deal to this work by adapting experimental psychology protocols to bumblebees that had previously been reserved for humans and a handful of other vertebrates considered to be intelligent, such as great apes and corvids. By observing bumblebees foraging on plastic flowers in a small closed box with a transparent lid, Chittka and his team demonstrated that the bees were able to learn from each other through simple observation.[13] The experiment consisted of teaching an "observer" bumblebee to forage on flowers of a certain color (blue flowers) by observing a "demonstrator" bumblebee that was already familiar with the task. The demonstrator bee was first trained alone, learning by trial and error that blue flowers contain sugar water and that flowers of another color (green flowers) should be avoided because they contain water mixed with quinine (a bitter substance that these animals don't like and that I advise you not to taste either). After several attempts, this demonstrator bee becomes experienced and makes very few mistakes when given the choice between blue and green flowers. Next an observer bee that has never seen the flowers is placed behind a transparent screen to learn the task. When the observer is then tested alone in the enclosure, it goes to the blue flowers on its first attempt, showing that it learned by simple observation.

And that's not all. Social learning can also occur between observers and demonstrators of different species, such as bumblebees and western honeybees. This demonstrates how important social information is for pollinators, regardless of where this information comes from. Indeed, it's common to see bumblebees foraging on the same flowers as western honeybees. The presence of a pollinator on a flower is an accurate signal providing information about the amount of nectar or pollen available in the flower. Surprisingly, and contrary to what has long been believed, these social learning abilities are not specific to social species. Some strictly solitary butterflies learn to recognize the most nectar-rich flowers by observing bees. Even more surprisingly, the information transfer in bumblebees works even if the demonstrator bees are replaced by resin models of their likeness, such as pointed ovals painted yellow and black or other colors. This shows that learning by observation involves classic forms of associative learning between visual stimuli and reward, without the need for new social-related cognitive abilities.[14]

Through this same process of observational learning, bumblebees are able to acquire new foraging techniques. Once again drawing upon experimental psychology research, Chittka and his team established that bumblebees are capable of learning to pull a string to obtain a reward, a behavior that does not exist in nature.[15] Even though bumblebees are often required to push, pull, and shift flower petals to maneuver into a flower's corolla, flowers are not equipped with strings. But, with a little patience and imagination, it was possible to make a demonstrator bumblebee understand that it could pull a string to access a drop of sugar water deposited on an inaccessible disk visible under a plexiglass table. An observer bumblebee watched the demonstrator bee pull the string and was then able to perform the task itself when placed alone in front of the string. This behavior spread from bee to bee, gradually forming a transmission chain, to such an extent that it was considered a culture-like phenomenon.

Shattering the myth of insect as automaton, the same team showed that bumblebees were able to improve on an observed behavior, a feat comparable to our ability to innovate on the basis of common knowledge. This time, Chittka and his colleagues showed bumblebees how to move a small ball into a hole to get a drop of sugar water.[16] I was lucky enough to witness this experiment. By the way, I still can't decide if the most accurate description is that the bumblebees learned to play golf or soccer. (If several bumblebees had been trained at the same time in the same enclosure, then without hesitation I'd say that they were playing rugby because bumblebees don't hesitate to throw elbows in a crowd for a drop of nectar.) Here again, a behavior that doesn't exist in nature was able to be transmitted from one bee to another by simple observation. But the bee's prowess didn't stop there. If the demonstrator bee's behavior was deemed ineffective by the observer bee, the observer would improve upon it. For example, if the demonstrator was taught to move the ball farthest from the hole when given a choice of balls placed at different distances, the observer would choose the ball closest to the hole instead, thus reducing the distance traveled, rather than copy the demonstrator without thinking. In this instance, we see the bees' culture evolving.

Beyond the astonishing and sometimes amusing nature of this type of discovery, we might wonder if we're really describing a similar phenomenon to the one seen in humans. Or is it something totally different that we concluded too quickly would never occur in nature? The answer to these questions is often right in front of us. In this particular case, we can find the answer simply by watching bumblebees forage in late summer. At that time of year, it's common to see bumblebees visiting flowers but not by landing on them or by entering them. Rather, they cling on to them from below by grabbing hold of a flower stem. Then the bumblebees introduce their proboscis by piercing the base of the corolla. By doing so, they obtain large quantities of nectar with less effort. They don't deposit or collect pollen, however, swindling the flowers of the usual contribution to their reproduction through pollination. This opportunistic behavior, sometimes referred to as "robbing," emerges randomly and then spreads through populations. It's sufficient for a single bumblebee to understand that it's possible to make a hole at the base of a flower with its mandibles and that from this hole more nectar will flow than from the flower's nectary. This innovative bumblebee's tendency will then be to continue puncturing the stems of other flowers of the same species, but it may also reuse the holes it already made to feed on the flowers it already visited. This bee is the primary robber. Other bumblebees, having observed all this, may start to make holes as well or will use existing holes. This makes them secondary robbers. This nectar robbing behavior observed in bumblebees can also be transmitted to other pollinators, notably western honeybees. Fortunately for plants, the latter, which are much more numerous than bumblebees,* are incapable of making holes by themselves because their mandibles are too small. Western honeybees use the holes already created by bumblebees. They're excellent secondary robbers. Every year at the same time, miniature cultures appear and then disappear as the bumblebee colonies develop in the spring and collapse in the fall, once the future queens disperse.†

These cultural transmission phenomena are probably much more widespread in insects than we think. For example, Étienne Danchin and Guillaume Isabel, my colleagues at the University of Toulouse in France, have revealed the existence of culture in *Drosophila* flies,[17] whose adult brain has two-thirds fewer neurons than that of a bee.‡ *Drosophila* are those little flies that hover around fruit baskets and beer glasses, and whose maggots develop in only three days. Their ultrafast life cycle (about ten days) and the ease of maintaining them for breeding have made them one of the most common animals in laboratories around the world. Today fruit flies

* Mature bumblebee colonies may have between two hundred and five hundred workers, while western honeybee colonies may have between forty thousand and eighty thousand workers.

† Bumblebees have an annual colony cycle. The queen founds the colony in spring. She produces workers until the end of the summer. Later in the season, the future queens and males mate and leave the nest to hibernate during the winter. The queen dies, and the colony disappears.

‡ In the fruit fly (*Drosophila melanogaster*), the larval brain contains one hundred thousand neurons, and the adult brain has three hundred thousand.

are probably the most studied organism, along with humans and mice. Flies have made possible countless discoveries on the functioning of living things, thanks to genetic manipulation that makes it easy to produce mutant populations that express specific behaviors.* Paradoxically however, very little is known about the behavior of these flies in the wild. Prior to Danchin and Isabel's research, other teams had pointed out that *Drosophila* tend to cluster on their favorite food sources—namely, slightly overripe fruit that has begun to decay. There on the fruit, ritualized sexual parades take place during which the females appear to select a male. A male, who perceives a female by her scent, would then lick the female's genitals, move closer to her so as to form a forty-five-degree angle with both their bodies, spread his wings, and produce a song inaudible to our ears. The female would either refuse the male with a rejection sound or accept him by letting him climb on top of her to proceed with mating. I'll spare you further details.

Danchin, Isabel, and their team have shown that during these gatherings, females were able to form a preference for a particular type of male by observing the choices of other females.[18] The experiment involved dusting males with colored powders. As a result, after seeing a female demonstrator mate with a male of a given color (for example, green), a female observer fly was highly likely to mate with a male of the same color, to the detriment of the competitor male (pink). This behavior can be maintained in a population of flies for several generations by being transmitted from the oldest to the youngest, even if it's not followed by all the females in the population. To demonstrate this maintenance of behavior over several

* Neurogenetics studies the role of genetics in the function of the nervous system. It emerged from the pioneering work of Seymour Benzer on *Drosophila* flies in the 1960s. Benzer (1921–2007) became interested in the link between circadian rhythms and genes, which led him to study other behavioral traits and neurological disorders using innovative techniques to express or inhibit at will certain genes coded for target proteins or involved in behaviors of interest.

generations, a female observer was placed in an enclosure where she could watch through glass windows as six couples formed and mated around her. A reverse "peep show." In most cases, female observers would follow the dominant behavior when given a choice between two males of each color. These artificially colored green males therefore enjoyed an artificially generated popularity rating for several generations. Like us, *Drosophila* follow fashion trends, and they're sometimes irrational.

A Miniature Theory of Mind?

The capacity to recognize individuals and recall past interactions with them promotes the emergence of complex social behaviors, such as coalitions and reciprocity. As human beings, we possess sophisticated forms of social intelligence that allow us to attribute unobservable mental states—such as feelings, desires, and beliefs—to ourselves or to other individuals. The capacity to understand the intentions of others is essential in our interactions. This is called "theory of mind." In children, this capacity is tested by asking them to solve the following task: "Max and his mom are in the kitchen. They put the chocolate away in the refrigerator. Max leaves to play with his friends. While he's gone, his mom decides to bake a cake. She takes the chocolate from the refrigerator, uses some of it, and puts the rest in the cupboard. Later, Max returns home. He wants to eat chocolate. Where will he look for it?" To answer correctly that Max will look for the chocolate in the refrigerator based on the information available to him, we need to attribute a false belief to him about where the chocolate is. The ability to do this develops in humans around age four or five.

Beginning in the 1970s, some great apes, dogs, and corvids were found to have the ability to attribute affective and cognitive states to other individuals based on their emotional expressions, attitudes, or presumed knowledge of reality.[19] The discovery gained consensus when ethologists found that chimpanzees could lie and cheat to divert others' attention.[20] In animals that belong to groups with hierarchies, some individuals temporarily hide food from other dominant individuals so that it won't be stolen. Others go so far as to simulate the presence of a predator by emitting alarm calls and waiting until all the members of the group have found shelter before leisurely savoring the food left behind.

While it's becoming increasingly clear that bees and other insects are capable of acquiring information by observing their congeners, to my knowledge, a theory of mind has never been studied in an insect. The very idea that these miniature beings might have such a skill has seemed far-fetched. But a recent study by Liz Tibbetts and her team at the University of Michigan suggests that the idea is not so impossible after all.[21] These researchers made certain wasps (*Polistes fuscatus*), which are capable of establishing hierarchies based on the recognition of facial markings, observe other wasps fighting. When the observer wasps were then introduced to one of the demonstrator wasps for their turn to fight, the observers adjusted their level of aggression depending on the demonstrator's aggression during the observed fight. Therefore, the observers were less aggressive toward the wasps they had seen being more aggressive and receiving less aggression, probably because they assessed that their chance of prevailing against these aggressive wasps would be low. Conversely, the observers were more aggressive toward opponents who had lost their previous fight. This behavior was not seen if the observer wasp encountered a fellow wasp that it had never observed before. Wasps are therefore able to recognize individuals in the group, assign them a social rank, remember observed interaction networks (who beat whom), and infer their own chances of winning a fight. But this experiment alone doesn't confirm the existence of a theory of mind in these insects that is comparable to that in humans and chimpanzees.

Given the considerable progress made in understanding insects' social intelligence, it's nevertheless quite possible that a theory of mind hasn't been discovered yet because the right protocol for testing it hasn't been identified. Perhaps we should turn once again to science fiction for inspiration? I recently rewatched *Starship Troopers*, directed by Paul Verhoeven. In this film, humanity is at war with an aggressive alien species called the Arachnids. Despite their name, these creatures don't look anything like spiders. Rather they're more like huge insects, given their segmented bodies and six legs. Their behavior is also remarkably similar to that of ants and bees, since they also have a division of labor between workers, soldiers, and queens. These aliens have a complex biology; the queens feed on their prey's brains and remotely control the other members of the colony. Each worker seems to have a sophisticated social intelligence. You can see this just by the ease with which they anticipate the movements of their fellow creatures and their human prey during fights. There's no doubt that these social insects from outer space show all the abilities associated with a theory of mind. What if Verhoeven was right in attributing a theory of mind to insects, just as del Toro attributed social recognition to Dictyoptera in the very same year? It's likely that we'll get the answer to this question in the near future.

3

The Limits of a Miniature Intelligence

Our brain is the result of millions of years of evolution. It may be large and powerful, but this organ we inherited from distant ancestors has limited capacities, as is the case with our other organs, such as our eyes and lungs. Unlike a computer that engineers can constantly upgrade with new components as technology advances, it's not possible to update our brain circuits. While it's true that our brain evolves, it does so slowly and according to strict biological and physical constraints that we can't overcome. Despite its sophistication, the human brain isn't perfect.

A healthy adult brain can frequently suffer from varying degrees of cognitive impairment. For example, there are many people who, like me, never remember where they left their keys. After our annoyance passes, we often end up finding them through logical reasoning, such as by checking our pockets or the key basket by the front door.

Sometimes these memory lapses are more inconvenient. Several years ago, one lapse in particular even cost me a car—or so I thought. Over the span of a few months, my car was stolen three times (I probably didn't park in the safest places). On the third occasion, the thieves stole the car my mother had lent me to replace mine. When I realized that her car was no longer at my house, I reported the theft to the police and the insurance company, as I had done the two previous times. Several weeks later, as I was walking to the laboratory one Sunday, I found the car in the middle of the empty university parking lot. At first I thought the thieves were playing a trick on me. Then I realized that my episodic memory, our memory that allows us to associate events with specific places and times, had failed me. I have no recollection of ever having parked the car there. And yet, there it was. It's possible to park our car as we would set down our keys: mechanically, without paying attention. But when something interferes with this routine, perhaps because there are no more parking spaces in the usual lot and we have to park in a new place, we might not remember the event. Or rather we never recorded it.

Bees, too, are endowed with learning and memory capacities, thanks to their brain. Apart from the fact that this brain is tiny, it shares many similarities with our own. But come to think of it, you've probably never seen a bee's brain. To better understand what follows, I'll describe it to you.

Imagine that you find yourself face-to-face with a bee and that this bee is very large (this detail is important, otherwise you won't see anything). You look the bee straight in the eye. If you concentrate hard enough, you can see through the hairs that cover its head (the scopae) and then through its cuticle (the equivalent of our skull). As you continue to concentrate, the bee's head becomes completely transparent, and its brain appears. Seen from the front, this brain resembles a large butterfly with its wings spread, occupying the entire volume of the head. Imagine this butterfly has spots (called "eyespots") on each of its upper and lower wings, like the peacock butterfly* commonly found in gardens in spring. In the bee's brain, there are paired structures (the butterfly wings) and un-

* This medium-sized butterfly (wingspan of 5 to 6 centimeters) is common throughout Europe and easily identifiable by its bright eyespots on a rusty-red background—a pattern reminiscent of peacock feathers.

paired structures (the butterfly body). Odors are perceived by the bee's antennae and then processed by the antennal lobes (the eyespots on the lower wings), their olfactory brain area. Images are perceived by the bee's compound eyes and then analyzed by the optic lobes (the upper wings).

These structures at the periphery of the bee's brain are involved in basic mental operations, such as learning to associate a stimulus with a reward. They exchange information with other more central brain areas through neurons (the veins on the butterfly wings). The neurons communicate with each other at the synapses, where chemical reactions take place. A bee brain contains about one million neurons and one billion synapses. Some neurons have specific functions, such as the reward neuron (VUMmx1)* that activates whenever the bee tastes sugar.[1] This particularly long neuron communicates with most other areas of the brain. But generally, neurons are organized in more complex circuits. This is true of the neurons that connect the optic and antennal lobes with the central areas of the brain, which are called "mushroom bodies"† because they resemble two large porcini mushrooms planted in the middle of the brain (the eyespots on the upper wings of the butterfly). Each of these central structures is made up of approximately 170,000 neurons that form a calyx. It's here that visual and olfactory information is integrated and that the most sophisticated forms of learning occur.

As in our brain, these areas specialized in information processing send orders to the rest of the body to produce behaviors. This information passes through the bee's body via a series of thoracic and abdominal ganglia (you can't see them because they're located behind the brain that you're still looking at from the front).

* This neuron was named the "ventral impaired median neuron of the maxillary neuromere 1" because of its location in the brain.

† We owe the description of mushroom bodies (*corpora pedunculata*) to French biologist Félix Dujardin (1801–1860).

These orders can be disrupted by drugs that block or activate certain neural networks in a targeted manner. For example, injecting a drug into the area of the brain that processes movement-related information, the central complex (the thorax of the butterfly), can induce paralysis. This is the mechanism some parasitoid wasps use to neutralize their prey while keeping them alive. These wasps inject a neurotoxic cocktail into the brains of cockroaches, "zombifying" them.[2] "Zombie" cockroaches are still able to move, but they're rendered completely passive, so that the wasps can guide them without difficulty to the wasp nest by pulling their antennae; the wasps then lay eggs inside them.

For a little over a century, research on bee behavior and their brain has revealed a surprisingly rich cognitive repertoire, previously unexpected in insects. The more we look for sophisticated cognitive abilities in their miniature brains, the more we find. In fact, we have uncovered so much in terms of their intelligence that we're now wondering what the limits of these creatures' intelligence are. Bees, too, have an episodic memory that allows them to associate events with places and times. For example, they're able to learn to forage on artificial flowers of one color in the morning and a different color in the afternoon.[3] But would they lose their car if they had one?

On The Shoulders of Turner

The study of insect intelligence has a long history. Karl von Frisch is generally credited as the pioneer in this field because of his experiments on bees.* Before him, famous naturalists such as Charles Darwin and Jean-Henri Fabre were considered avant-garde for investigating the existence of intelligent behaviors and mental ca-

* See chapter 1, A Poor Sense of Direction.

pacities in insects. But their conclusions were based on inferences from observations and rarely from experiment results. Other researchers, such as Sir John Lubbock (1834–1913),* conducted extremely creative experiments, but these sometimes lacked the scientific rigor necessary to draw firm conclusions. At that time, since it wasn't possible to explore the intricacies of their intelligence, we were mainly trying to understand how insects perceived their world. For example, in 1883 Lubbock conducted experiments to test whether western honeybees could communicate with sound.[4] He connected two groups of bees by placing a telephone handset with a microphone above each group of bees (his friend Alexander Graham Bell, the inventor of the telephone, helped with this setup). When Lubbock disturbed the bees at one end of the line, the bees at the other end weren't affected. He concluded that bees do not use acoustic communication. Unfortunately, Lubbock didn't continue his experiments by subjecting the bees to other sound frequencies or by observing them in other contexts. If he had, he'd probably have discovered that acoustic communication does play a major role in bees, such as during swarming, for example.

It wasn't until the beginning of the 20th century, in the infancy of ethology, that Karl von Frisch documented the first series of convincing experiments on bee cognition, well before he translated the meaning of the bee's waggle dance in the 1940s. In 1915, von Frisch demonstrated that bees have color vision.[5] To do this, he trained western honeybees to forage on small colored cardboard platforms placed on a table. The insects had to learn to associate a color (blue, for example) with a drop of sugar water placed by

* A famous British naturalist, prehistorian, banker, and politician, Lubbock was introduced to biological experimentation by his neighbor and friend Charles Darwin. His experiments on the behavior of Hymenoptera are rare examples of creativity and an inspiration to many ethologists. In the tradition of Darwin, Lubbock's research sought to raise the intellectual rank of insects to be close to that of humans.

pipette on the piece of cardboard. Platforms of other colors offered an unsweetened drop of water and were therefore not attractive. To verify that the bees were using the colors and not just spatial cues, von Frisch shuffled the platforms on the table each time a bee returned to visit the setup. He observed that the foragers that had found the sugar water on the blue platform tended to return to that same platform on subsequent visits, even if it had been moved in the meantime. To make sure that the bees weren't simply using the contrasts in brightness (light intensity) to solve the task, he hid the blue piece of cardboard among pieces of cardboard tinted with different shades of gray. While some of these gray cardboard pieces had the same light intensity as the blue piece, the bees were never tricked. This meant that they could see the colors. With this approach, von Frisch was able to demonstrate that bees could discriminate a range of colors from violet to orange but that they couldn't differentiate red and gray. Bees are sensitive to a slightly different spectrum than we are, which allows them to excel at recognizing flowers of different species.

It has been a little more than a century since insect intelligence began to be studied through rigorous scientific experiments. But in looking at these first studies, we cannot disregard the contribution of Charles Turner (1867–1923), another pioneer in the study of animal cognition, whose work is paradoxically as remarkable as it is unknown.[6] In the wake of the Black Lives Matter* movement, which has recently gained international attention, the scientific community is working to revitalize Turner's unjustly forgotten work one hundred years after his death. Here's why.

* Black Lives Matter was founded in 2013 in the US with the mission to eradicate systemic racism against the Black community. The organization gained international momentum during the summer 2020 protests of George Floyd's murder by police during his arrest.

Turner was an African American scholar, born two years after the abolition of slavery in the United States. At the turn of the 20th century, he conducted a series of revolutionary experiments in animal psychology that went against mainstream ideas of the time, which largely rejected the notion of advanced intelligence in any species other than humans. For example, in 1892, he published a study on the variability in spiderweb-building behavior. Contrary to the view that this behavior was instinctual, even robotic, Turner demonstrated the existence of significant variations in the geometry of spiderwebs made by spiders of the same species. Not all spiders followed the same strict construction plan. This observation was a precursor to the concept of animal personality,* which was theorized in the early 2000s and is now a field of study in its own right.

Much of Turner's research has been entirely forgotten, leading some of the most illustrious researchers to reinvent an already well-wrought wheel. In 1908, Turner documented how solitary bees used visual cues to locate their nests in the soil.[7] He noticed that placing an object (a Coca-Cola bottle cap) near the nest and then moving it a few centimeters caused the bee to look for the entrance to its nest in the wrong place, toward the displaced object rather than at the actual entrance to the nest. This experiment is similar to Nobel Prize winner Niko Tinbergen's experiment that I discussed earlier† in which an insect's nest was surrounded by pinecones and then the cones were moved around to trigger search behavior. Today Tinbergen's study can be found in all ethology textbooks, whereas Turner's study, published twenty years earlier, is practically unknown.

Similarly, in 1910 Turner affirmed that bees are able to associate a sugar-water reward with a color stimulus.[8] He constructed

* This concept refers to behavioral differences between individuals that persist over time. This idea is described in detail in chapter 4, The Superorganism.

† This experiment is described in chapter 1, A Poor Sense of Direction.

artificial flowers (using colored cardboard disks) containing a drop of honey and placed them in the middle of a bush to attract bees. By manipulating the shape, smell, and color of these cardboard flowers, Turner came to the same conclusion but five years earlier than von Frisch (who was also a Nobel Prize winner, like Tinbergen): namely, that bees can associate a color with a reward. Unfortunately, Turner chose to use red flowers that probably appeared gray to the bees, and he didn't perform the crucial experiment that would've ruled out the possibility that bees rely on contrast (shades of gray) instead of color. His experiment was therefore incomplete. Nevertheless, his contribution remains significant and unrecognized, as it was probably the first study to consider the foraging behavior of bees in the context of associative learning, establishing an association between stimuli and reward.[9]

In the end, Turner published seventy-one studies, some of them in leading scientific journals. In 1892, two years before earning his doctorate, he published two articles in the prestigious journal *Science*.* Had he lived in our time, this early success would've earned him a tenured research position at one of the world's top research institutes. In the late 1800s, however, his success wasn't enough to open the doors to even a moderately reputable American university. Turner thus pursued a career as a natural-science teacher at a high school for Black students and conducted most of his experiments on insects in his spare time. We can only imagine how much he would've accomplished if he'd had the same opportunities as his white colleagues at universities, with time and money to devote to his research, access to a university library, a team to help him, and a network of students and collaborators to make his work known.

* Founded in 1880, this American weekly publishes articles in all fields of science and is the world's best-selling and most widely read peer-reviewed general scientific journal, giving it a high impact factor.

This example is unfortunately not an isolated case, and it reminds us of another avant-garde researcher in the field of cognitive science whose work has recently resurfaced: the British mathematician Alan Turing (1912–1954). Persecuted during his lifetime for his homosexuality, Turing is now considered one of the most influential researchers of his time for his pioneering work in computer science and artificial intelligence. Fifty-nine years after his death, he was even recognized by the Queen of England as a national hero for having played a key role in decrypting the Nazi army's communications during World War II. Fortunately, times change, and diversity is increasingly becoming a part of all aspects of our society, including in research laboratories. As the examples of Turner's work on animal intelligence and Turing's on artificial intelligence show, diversity increases the talent pool and has the potential to transform entire disciplinary fields.

Flower Power

Influenced by these pioneers in animal cognition, the number of studies on the intelligence of bees has increased steadily and now indicates that these insects excel in a wide range of forms of learning. This is especially true for learning odors, which is essential for bees' communication and foraging. Olfactory memories allow bees to recognize each other* and to navigate over long distances to find abundant food sources, such as flowering meadows and trees, and to choose which flowers to visit once there. Plants of different species emit distinctive fragrances that bees can perceive from several yards away. But the bees themselves leave chemical fingerprints on the flower petals every time they land on them.[10] These scent marks are mainly composed of cuticular hydrocarbons (the same as those

* See chapter 2, The Fragrance of Déjà Vu.

that form the colony odor) and are therefore perceived from within only a few centimeters. They tell bees approaching a flower whether another bee has recently visited it, without the bee having to land on it.

All these odors, whether emitted by flowers or by bees, are first perceived by the bee's antennae and then processed by the olfactory system's neuronal circuitry in the brain to allow the bee to learn and form memories. The study of this learning process was greatly improved by the development of an innovative conditioning protocol in 1961 by Kimihisa Takeda at the University of Tokyo in Japan.[11] The protocol is like the famous experiments of Ivan Pavlov (1849–1936), the Russian physiologist who won the Nobel Prize in 1904 for his work on dog digestion, but adapted for bees. In his time, Pavlov noticed that presenting a dog with a piece of food (meat) was enough to make the dog salivate. For the animal, the piece of meat was an "unconditioned stimulus," meaning a signal that automatically triggered a salivation reflex, regardless of the context. Pavlov used this reflex to test whether dogs could learn associations between two stimuli: an unconditioned stimulus (meat) and a neutral sound that had no particular meaning for the dog. Pavlov started making a bell sound every time he served the dog its meal, and in a short amount of time the dog began to salivate as soon as it heard the bell, even when Pavlov no longer presented the dog with meat. The dog had learned the association between the bell and the meat, and salivated in anticipation. After this conditioning, the sound had become a conditioned stimulus, since it was able to induce a response even in the absence of the unconditioned stimulus (the meat).

Following the same principle, Takeda succeeded in conditioning western honeybees to associate a scent (the conditioned stimulus) with a reward of sugar water (the unconditioned stimulus).[12] When the antennae of a hungry bee sense sugar water, the bee's reflex is to extend its proboscis to suck up the reward. When the antennae detect a scent, if the bee is not trained, this proboscis reflex is not generated. If the scent is presented at the same time as the reward, however, the bee forms an association that subsequently allows the scent alone to trigger proboscis extension. Like Pavlov's dog, Takeda's bee can be conditioned by repetition. But, contrary to dogs, bees aren't docile animals. It's therefore necessary to keep a bee in a small custom-built tube so that it can't escape or attack during conditioning, while still ensuring that it can move its antennae and extend its proboscis. I've always admired my colleagues who prepare their bees for later testing by placing them in tubes and setting them on racks by the dozen. Surprisingly, the bees cope with this quite well, sometimes even giving the impression that the rows of antennae are standing at attention waiting to be tested.

This Pavlov-inspired conditioning protocol may seem strange. But its merit is that it reproduces on demand the behavior of a bee that has landed on a flower and decides whether to extend its proboscis to suck in nectar, while precisely controlling the scents the bee perceives, the time it's exposed to these scents, the number of exposures, and the perceived reward. It's therefore useful for exploring the capabilities of a bee's brain. Some laboratories have even specialized in this protocol by equipping themselves with automated devices that use dozens of odor cannons connected to computers. This way, these studies can be conducted like an assembly line, with bees moving through an experiment one after the other. With this conditioning protocol, a bee can be taught to associate a single scent with a single reward (this is called "absolute conditioning"). With a little more work, the same bee can be taught to discriminate between stimuli by learning that one scent is associated with a reward while another scent is not (called "differential conditioning"). The bee can then be taught the opposite task (reversal learning). Immobilized in the tube, the bees are also able to understand stimulus patterns, for example when two scents are individually associated with a reward but not associated in combination (negative patterning) or the reverse (positive patterning).

After this conditioning process, the bees keep the information in their memory for varying lengths of time. A forager bee can recall the association between a scent and sugar water for at least three days, even if it has only been exposed to the scent once in its life.[13] But the more these exposures are repeated, the longer and more robust the memory. Randolf Menzel and his team at Freie Universität Berlin studied where and how these memories form in a bee's brain by recording the neuronal activity before and after such training.[14] Bees are too small and too unruly to be placed inside a magnetic resonance imaging (MRI) scanner as a human patient and some other large animals can be. Nevertheless, it's possible to obtain images of a bee's brain activity by injecting a small amount of dye directly into its head (this operation is painless because the cuticle that covers the head doesn't have nerve endings). The dye reacts when calcium is emitted by active neurons. When an odorant is delivered to the bee, it stimulates specific neurons and produces a stain in the brain from the reaction of the dye with calcium. Calcium imaging has revealed that bees' antennal lobes are composed of masses of neurons (glomeruli) that produce a colored stain specific to each perceived odor. For example, imagine that the scent of mint produces a straw-shaped stain and the scent of ethanol produces a drinking-glass-shaped stain. When mint and ethanol odors are presented simultaneously to a bee, the activation pattern is the sum of the activation patterns of the two odorants presented individually: a mojito in a glass!

This associative learning works well with natural odors, such as certain pheromone compounds and flower fragrances, and with synthetic odors that bees are unlikely to encounter in nature. In fact, some researchers use bees' ability to form specific, long-term olfactory memories to detect odors that humans cannot sense, for example the odor of cannabis and traces of explosives. Bees can even detect COVID-19 in patients because, as with other illnesses, SARS-CoV-2 induces metabolic changes that produce characteristic body odors. So bees can be trained to use their sense of smell similarly to how we use sniffer dogs. This practice is not widespread, however, because it's extremely costly in terms of bee numbers. Unlike a dog, a bee has a life expectancy of only a few days. Just think how many bees would be needed for a single demining operation!

More elaborate conditioning that involves associations with multiple stimuli requires deep brain regions such as the mushroom bodies. Jean-Marc Devaud and his team in my laboratory in Toulouse have shown that blocking these brain structures by injecting a drug (procaine) prevents any learning that involves ambiguity, such as positive or negative patterning.[15] Bees' mushroom bodies are therefore the equivalent of mammals' hippocampus, a brain region known to integrate information of different kinds to form spatial and episodic memories. In western honeybees, these structures grow throughout adulthood with the establishment of new neuronal connections. These changes in brain morphology are thought to reflect bees' need to undertake increasingly more complex visual and spatial learning as they become foragers near the end of their lives. Similar brain plasticity occurs every day in our brains. Neuronal connections are formed when we learn new things, whereas the connections that are no longer useful to us disappear. This phenomenon can be seen when comparing populations with different professional activities. For example, London cab drivers have a posterior hippocampus (the area specialized in spatial memories) that is much more developed than average because they have to learn to navigate between hundreds of locations and therefore create complex spatial memories, which isn't the case for the majority of people.[16] The same is true when comparing the mushroom bodies of foraging bees and nurse bees that remain in the colony.

Concepts and Arithmetic

Bees are equipped with powerful eyesight that allows them to orient themselves using the landscape and recognize the flowers on which to land. Historically, their visual learning has been studied by making use of forager bees' tendency to make regular trips back and forth between their nest and a food source. In these studies, bees aren't kept in enclosures; they're free to move around in their natural habitat as they wish. When bees come to forage on a food source, researchers can then condition these foragers to associate images (the conditioned stimuli) with a reward of sugar water (the unconditioned stimulus). And so these bees learn that they have to fly to the stimulus to get the reward. This is "operant" conditioning, a form of conditioning theorized by the American psychologist B. F. Skinner (1904–1990) following his experiments on mice, which were strongly influenced by the work of Pavlov. In operant conditioning, bees can learn to differentiate colors or shapes that are characteristic of flowers. These associations are acquired over a period of several days or even several weeks, providing the opportunity to study more complex manifestations of learning and memory than when bees are restrained in tubes in the laboratory.

Under these free-flying conditions, bees are able to learn to follow logical rules, as some primates and birds do. Randolf Menzel of Freie Universität Berlin, Mandyam Srinivasan, who was then at the University of Canberra in Australia, and their teams have shown the ability of western honeybees to learn to systematically move toward stimuli similar to those seen on their route, which is akin to understanding the concept of sameness.[17] This research team tested western honeybees in Y-shaped mazes in which color images were presented at the maze entrance and at the end of each arm. To learn the concept of sameness, the bee had to understand that to get a reward, it had to choose the one of two images at the ends of the arms that matched the image shown at the entrance to the maze, which was replaced by a new image on each visit to the maze. The trained bees showed they had indeed learned a concept, not simply because they recognized an image, but because they were able to transfer the rules to a new stimuli, such as when images were swapped for odors. Similarly, these researchers were able to show that bees are able to learn the concept of difference. In the study, bees would systematically choose the arm of the maze containing the image or the odor that was different from the one presented at the entrance.

Other studies suggest that bees can even understand arithmetic concepts, similar to counting. When foraging, bees theoretically have many opportunities to count things, whether that's the petals of the flowers they land on or the visual cues along a route. Of course, it's unlikely that bees can recognize numbers. They do seem to have some ability to estimate and sequence quantities, however. Adrian Dyer and his team at the University of Melbourne in Australia conducted experiments that suggest bees can learn the concept of zero.[18] To do this, they trained western honeybees to choose between two images consisting of several black symbols on a white background in a two-branch maze. By systematically rewarding the bees when they chose the branch containing the visual stimulus with the fewest symbols, the researchers observed that the bees could extrapolate the concept of "smaller than" until they reached zero, when the image did not include a symbol (a blank sheet).

Lars Chittka and Karl Geiger, then at Freie Universität Berlin, explored the question of bees' counting ability in the more general context of navigation with a new hypothesis that bees might count visual cues to reach a target.[19] Chittka and Geiger placed large colorful tents along the route between a hive of western honeybees and an artificial flower. The tents, which were over 11 feet high, served as visual cues for the insects. To prevent other environmental elements from interfering with the experiment, the researchers selected a large field without trees and with a horizon line that was as flat as possible for the experiment site. Once the bees were trained to travel the route between the hive and the flower placed after the fourth tent, Chittka and Geiger removed the flower and varied the number and position of tents along the learned route. When the bees encountered more tents than before between the hive and the flower, they tended to stop after passing the fourth tent, even if it was placed before the usual position of the flower. On the other hand, when the bees encountered fewer tents than before, they tended to fly farther and thus overshoot the usual position of the flower. This indicated that they were counting visual cues to locate the food source, or at least had a sense of quantity.

This research has inspired numerous studies on logical reasoning in miniature brains. One of the most recent studies was on bees' ability to make transitive inferences, which is being able to predict new relationships between objects based on known relationships between other objects. We do this every day when we need to classify objects or events using value judgments. For example, if you want to read a book about insects and the bookseller tells you that the book on bee intelligence is much more thorough than the one on ant psychology, which itself is of better quality than the one on fly sociology, then you'll probably be tempted to choose the book on bees. (And it seems that's actually what you did, and I thank you for it.) My colleagues Julie Bernard and Martin Giurfa in Toulouse have tested this transitive inference capacity in western honeybees.[20] In their version of the transitive inference test, Bernard and Giurfa trained each bee to differentiate between visual stimuli (black-and-white patterns) in a specific sequence in a Y-shaped maze. Choosing pattern A was rewarded with sugar water while choosing B was not, then B was rewarded while C was not, then C was rewarded but D was not, and finally D was rewarded but E was not. These adjacent pairs underlie the hierarchy A > B > C > D > E. Once the training was complete, Bernard and Giurfa tested the bees' ability to choose between a nonadjacent pair of patterns: B and D. The bees had seen both pattern B and pattern D before but never at the same time. During the training, B and D were equally rewarded or not rewarded, so pure association with a reward would mean the bees would be expected to choose randomly between B and D. If the bees were able to create a mental representation of the hierarchy implied during training, however, they would prefer B to D. In this study, the majority of the bees failed to choose B, which would have indicated their capacity for transitive inference, because B is better than C (B > C), which is better than D (C > D), so B is better than D (B > D).

Conversely, *Polistes* wasps, which form dominance hierarchies to decide who will reproduce and who will care for the larvae,* passed the test.[21] But the experiment was done a little differently. Liz Tibbetts and her team at the University of Michigan trained wasps to differentiate between five visual stimuli (this time it was colored pieces of paper) in a two-compartment box with a metal floor. When a wasp went to the compartment that contained the "punished" color, it received a slight electric shock from the floor. Each wasp had to learn to differentiate between pairs of colors to avoid electric shocks; then the wasp was tested with a new pair of familiar colors, but ones that it had never seen together at the same time. Faced with this new choice, all the wasps were able to choose the color that did not predict the electric shock based on the learned

* See chapter 2, The Fragrance of Déjà Vu.

relationships between the colors. This difference between bees and wasps does follow a certain logic. It's likely that bees don't have the capacity for transitive inference because they don't need it. When foraging, bees specialize in one type of flower and only change when that type is exhausted in their environment. The ability to make transitive inferences is therefore not useful to them. On the other hand, wasps need to be able to make transitive inferences so they can know the hierarchy between dominant and subordinate individuals in colonies. Because, in these small colonies, it's crucial to know who you're dealing with!

Bumblebees Aren't Always in a Good Mood

Beyond their impressive capacities for learning, insects have a rich emotional life as well, as suggested by a growing body of research. What do bees feel when they're foraging? Are they happy to go out and explore the world? Are they afraid of encountering crab spiders?* Are they jealous of other bees that bring back more food than they do? This aspect of insect cognition is much more complex to demonstrate than learning because bees can't tell us about their state of mind. We only have access to their behavior and sometimes to their physiological measurements, which we must interpret without ever being sure of what they mean. This research is therefore largely subject to uncertainty and scientific debate.

As social animals, we're well aware of the importance of interactions with others for our well-being. Isolation is a source of stress that affects our mood and our health. A few years ago, I wondered

* A group of mimetic spiders that are ambush predators. They are called "crab spiders" because of their long pairs of front legs that give them a crustacean appearance. They're among the main predators of bees, which don't detect the spiders easily when landing on flowers.

whether insects would also suffer under imposed lockdown. It has long been known that when a juvenile cockroach (a larva) is isolated from its group for several days, it develops much more slowly and has altered social behaviors compared to a cockroach kept in a consistent social environment. With Colette Rivault, my cockroach-specialist colleague at the University of Rennes, I hypothesized that these developmental disorders were due to a form of emotional stress felt by cockroaches lacking physical contact with others. What if cockroaches are also prone to depression?

I started building a device that would provide on-demand social stimuli to insects placed in isolation: a "cockroach-petter."[22] At that time, laboratories were not equipped with 3D printers. If you weren't handy enough to make a device yourself, then you at least had to be creative. My cockroach-petter was a plastic box in which a rotating broom—really a goose feather—would provide a tactile stimuli every thirty seconds to simulate contact with another cockroach. I used a dental elastic connected by a system of pulleys to a disco-ball motor to make this broom move, and the entire setup was held together miraculously by adhesive putty. Equipped with one hundred of these motorized brooms, I compared the developmental rates of isolated cockroaches with no tactile stimuli with those of other cockroaches placed either in the presence of a stationary or a rotating feather. After one month of observations, the artificial social stimuli generated by the rotating feathers had markedly accelerated the cockroaches' development to levels comparable to those of control cockroaches left in their groups. I later demonstrated that this astounding effect could be recreated by putting a cockroach in a box with a bunch of moving objects. It also works well to confine a cockroach with crickets that jump on it regularly. These poor cockroaches . . . Was this physical contact enough to make them less "depressed"? We've never fully understood all the processes at work in these experiments. It's likely that the regular physical contact induced physiological responses in the cockroaches, such as the stimulation of juvenile growth hormone production. What is certain, however, is that this type of research has inspired many more experiments.

As a result, more recent studies have investigated the challenging subject of insect emotions. Chittka and his team at Queen Mary University of London explored the existence of positive and negative emotions in bumblebees.[23] Because of their round and hairy appearance, bumblebees are generally thought to be harmless. If I didn't take a moment to set the record straight on this misconception, I'd be disrespectful to my colleagues whose faces have sometimes doubled in size as a result of allergic reactions to bumblebee stings! Bumblebees do sting, and unlike western honeybees, which lose their stingers and die after stinging, bumblebees don't lose a thing. They can therefore sting several times in a row if they feel like it. Fortunately, they aren't too aggressive, and to be attacked you have to seek it out, for example by putting your hand in a hive (this is what my colleagues often do when they carry out experiments). Just like with western honeybees, it's the female bumblebees—the workers and the queen—that sting. Males don't have a stinger, but there aren't as many of them, and they look a lot like the workers. Therefore, you need to be relatively self-assured before greeting one of these insects with bare hands.

Based on the observation that bumblebees can sometimes (but rarely) get angry, Chittka and his team used an experimental-psychology protocol to explore the existence of emotions in these insects.[24] In this "judgment bias" test, the bumblebees had access to a small wooden enclosure connected to the nest by a pipe. The enclosure was covered with a transparent plexiglass roof so that the researchers could observe the bumblebees without them escaping. In this setup, the bumblebees had to learn that a blue platform up against one of the enclosure walls was associated with a sugar-water reward, while a green platform provided no reward. Once trained to differentiate these colors, the bumblebees were presented with a platform of intermediate color—not blue or green, but turquoise. The amount of time the insects spent visiting this ambiguous stimulus depended on what had happened immediately before the test. When the bumblebees were given an unanticipated drop of sugar water just before being released into the enclosure, the bumblebees rushed to the turquoise platform, interpreting it more like blue. The behavior of these bumblebees appeared optimistic: they saw the glass as half-full. On the contrary, bumblebees that received a light pressure on their body from a pair of pliers (as a crab spider's grip would feel to them) took much longer to visit the turquoise platform, interpreting it as green instead. The behavior of these bumblebees was pessimistic: they saw the glass as half-empty.

Another indication that emotional states vary in insects is the fact that some insects use psychotropic substances that affect their nervous systems—an option we might turn to when we're going through difficult times. Plants and mushrooms naturally produce anesthetic substances, appetite suppressants, depressants, and hallucinogens. These compounds allow the flora to defend themselves against herbivores. But they don't repel everyone. Bumblebees, for example, prefer to forage on flowers that have low levels of nicotine in their nectar rather than on flowers that don't contain any.[25] Nicotine seems to promote certain kinds of learning, such as the ability to differentiate between flowers of different colors. Therefore, producing nicotine-rich nectar can be seen as a manipulative strategy employed by a plant to retain bumblebees.

Alcohol—in the form of fermented fruits—is also widely present in nature. Galit Shohat-Ophir and her colleagues at Bar-Ilan University in Israel found that *Drosophila* flies would drink alcohol when stressed,[26] probably to forget the adverse circumstances they're experiencing. In these flies, mating stimulates the reward circuit in the brain, while sexual deprivation inhibits it. To test the effect of these brain signals on fly behavior, Shohat-Ophir created two groups of male flies. In the first group, over the course of several hours, males were allowed to mate with many unmated females. In the other group, males were placed with already mated females who systematically rejected their advances. When the male flies were then given a choice between food containing ethanol or no ethanol, only those deprived of mating consumed the alcohol. In the long term, this attraction to ethanol could have undesirable consequences, as the flies did show addictive behaviors, consuming more and more alcohol over time and disregarding aversive stimuli, such as quinine, in order to obtain the alcohol. Were they drowning their sorrows? Or compensating for their frustration? In any case, it seems that insects also have their ups and downs.

A Miniature Imagination

What do bees think about when they dream? This question brings up another intriguing aspect of the secret life of insects: their imagination. While bees can learn to recognize many objects, such as a hive, flowers, and different kinds of visual cues, this doesn't necessarily mean that they're able to build a mental image of these things. To test this, Cwyn Solvi and his colleagues at Queen Mary University of London investigated whether insects could learn to recognize objects by sight and differentiate them by touch.[27] The researchers presented bumblebees with artificial flowers of two different shapes—either cubes or spheres—in a lit enclosure. Only one flower shape contained a sugar-water reward. To ensure that the bumblebees were only learning by vision, the flowers were placed under a piece of transparent plexiglass so that they couldn't touch them. Then, when the bumblebees were moved into the dark, the researchers observed that they were able to differentiate between the cubes and the spheres based on touch alone. The reverse was also verified (learning in darkness and testing in the light), strongly suggesting that bumblebees develop a mental representation of an object's shape when they encounter it, instead of simply learning to recognize it by sight or touch. In our own everyday lives, we use this type of imagination when we recognize by touch objects that we've seen before, like when we're feeling for something on the top of a shelf or at the bottom of a bag. We don't see the objects, but we recognize them thanks to a mental representation that combines tactile, visual, and auditory information.

Looking beyond simple objects, Martin Egelhaaf and his team at Bielefeld University in Germany studied bumblebees' ability to form a mental representation of their own bodies.[28] Bumblebees in the same colony vary greatly in size. The queen can be up to ten times larger than the smallest workers. These morphological differences depend on the position of the cells in which the larvae develop in the nest. The closer the larvae are to the center of the nest, the more food they receive from the workers, and the warmer they are because the nest is compact and attracts more individuals to that area. These larvae become larger adults than those that hatch at the periphery of the nest. Egelhaaf and his team investigated how the bumblebees comprehended these variations in size when they had to pass through small holes, for example to maneuver through dense vegetation to get back to the nest. On a human scale, this is like judging whether our body will fit walking straight through a doorway or whether we'll need to turn sideways. Our ability to make this type of judgment is the result of a long learning process. To test whether bumblebees are capable of this judgment, the researchers had bumblebees of different sizes pass through a hole in a wall to return to their nest. They filmed the bumblebees' flights with high-resolution cameras. When played in slow-motion, these videos solved the mystery. When a bumblebee's wingspan was smaller than the diameter of the hole, it passed straight through the center. When its wingspan was larger than the hole, however, the bumblebees automatically flew at an angle to avoid crashing. Before passing through the hole, they each examined it in detail, from side to side, to judge its width and the probability of hitting the edges. Therefore, bumblebees have a mental representation of the shape and size of their body. Exactly how they acquire this mental representation isn't known, but one possibility is that they acquire this information while walking, such as when they're in their underground nest, because their step size correlates to their body size.

Another aspect of imagination is the ability to anticipate the consequences of our actions—or in other words, the ability to plan. Before making a decision, we usually try to estimate our chances of making the best choice. This awareness of our own cognitive processes is called "metacognition." Clint Perry and Andrew Barron of Macquarie University in Australia have explored the possibility that insects are capable of metacognition by putting them in ambiguous situations.[29] The researchers trained western honeybees in a two-chamber apparatus in which they were rewarded with sugar water when a visual stimulus (a colored shape) appeared above a black reference bar. When the colored shape appeared below the reference bar, however, the bees were punished with a quinine solution. After training the bees on the concepts of different shapes, sizes, and colors of visual stimuli, Perry and Barron presented them with ambiguous stimuli in which the shapes were no longer clearly above or below the reference bar, but roughly on the bar. The more difficult the task was to solve (because the stimuli could not be easily interpreted), the more the bees would choose to exit the apparatus, preferring not to make a choice rather than risk getting it wrong. This suggests that they were able to gauge their degree of uncertainty in solving a problem, thus demonstrating metacognition.

Bees likely use even more elaborate forms of planning in their daily lives. Western honeybees in particular are renowned for building nests out of wax, pollen, and propolis in an extremely uniform pattern. In a modern human-made hive with frames outfitted with wax foundations, the visible part of each honeycomb cell is a hexagon, with each side measuring 3 millimeters long, a depth of 11.5 millimeters, and a thickness of 0.05 millimeters. Each cell is joined to three others via a surface formed by three rhombuses on a plane tilted upward by 7 to 8 degrees to prevent the honey from draining out of the cells. This structure maximizes stability while minimizing building material, an architectural feat of perfection discovered prior to 300 BCE by Aristotle, and subsequently by numerous mathematicians and physicists. The bees' work gives the impression that it's the result of an innate behavior that leaves no room for improvisation. In the wild, however, western honeybee colonies are found in natural shelters that vary greatly in terms of volume and shape. Tree-trunk hollows are one such example that can constrain nest construction considerably. To verify bees' improvisational construction behavior, Kirstin Petersen and her team at Cornell University analyzed the characteristics of more than twelve thousand cells built by bees in different types of hives and natural shelters.[30] From this detailed analysis of nest architecture, the researchers found that workers adapt their building behavior to the geometric constraints of a space, sometimes building cells larger in size, of more irregular shape, or with varying degrees of wall tilt. It has been suggested that the diversity observed in honeycomb construction could be the result of a bee's mental-image template of the desired outcome when they begin construction.[31] Accordingly, the bee would perceive the existing geometry and decide to build an irregular hexagon—or even a pentagon or a heptagon—as a compromise in order to ensure the structure's stability. This architecture would allow the bees to alternate between constructing worker cells and drone or queen cells, which are much larger and protruding.

Consciousness and the Temptation to Anthropomorphize

On August 6, 1882, Sir John Lubbock conducted an experiment to compare the "taste for work" of wasps and bees. For a full day, he observed a wasp and a bee collecting honey on a table.[32] The wasp made 116 visits between 4 a.m. and 8 p.m., while the bee made only 29 visits. Lubbock concluded that wasps are more industrious than bees. This example from an earlier time illustrates the pitfalls of studying animal cognition. First, Lubbock constructed the question he was investigating in this experiment from a purely human perspective. What is an insect's taste for work? Is collecting honey to feed its larvae really comparable to work as we understand it? Second, Lubbock concluded that wasps are more industrious than bees without trying to rule out other hypotheses that might have explained this result. What if Lubbock had stumbled upon a particularly experienced wasp and a novice bee who was just learning to forage? Did these wasp and bee colonies have comparable food requirements? Or are wasps simply more tolerant of cool morning and evening temperatures than bees, allowing them to forage for longer?

For several decades now, ethological research has been pushing the frontiers of our knowledge about animal intelligence. In insects, this research is more recent, and the limits of insect intelligence may be far more vast than we'd thought they were. But, just as we can't make any conclusions on the industriousness of wasps and bees from Lubbock's experiment, many interpretations of other experiments are not definitively settled, since this research often borrows hypotheses and protocols based on human psychology. As humans, we're capable of carrying out cognitive tasks that we consider complex, and we want to test whether other animals are capable of solving the same tasks by adapting the protocols to different species. But does the question being tested make sense? At first glance, it's not all that interesting to test whether bumblebees know how to do a sudoku, unless observations suggest that the ability to handle the logic of numbers would be useful for their everyday life, foraging, social interaction, or nest building. The temptation to anthropomorphize is therefore significant, and many theories about the capacity for empathy, tool use, counting, and metacognition, to name but a few, are still widely debated because they require rigorous exclusion of all alternative hypotheses.

These debates, sometimes within the same laboratory, call upon researchers' practices and philosophical positions, which can be different from one another's. It's this confrontation of ideas—the very principle of scientific research—that incites experimentation in several directions and contributes to a better understanding of the phenomena being studied. Sometimes a debate results in qualifying insects' prowess by putting forth more parsimonious explanations.* For example, one study on spatial concept learning in bees looked beyond whether bees could solve the task to consider how they solved it. Detailed analysis of the flight paths of bees learning the "concept of aboveness" in a Y-shaped maze showed that some of the bees never looked at the entirety of the visual stimulus (the shapes below and above a horizontal bar) but used only part of it to locate the reward. From a close-up scan of only the bottom part of the stimulus, it was possible for the bees to solve the task without learning the concept, turning it into a simple discrimination task.

* In psychology, the principle of parsimony was formalized in 1903 by Lloyd Morgan, and it states that if simpler psychological processes can explain an action, then this explanation must be favored over more advanced psychological processes. This principle is also known as "Morgan's Canon."

The bee can behave as if it had learned the "concept of belowness" by systematically visiting the arm of the maze containing a stimulus below the horizontal bar regardless of whether there is anything above the bar or not.[33] At other times, a debate reveals an even higher level of intellectual sophistication. In trying to understand the mechanisms of social learning in bumblebees, it was discovered that observer bees were able to copy their peers and to improve upon observed behaviors.[34] Asking the right questions and finding the right protocols to rule out alternative hypotheses is what comparative psychology is all about.

At the moment, the most debated question concerns the existence of consciousness in insects. In humans, the term "consciousness" is generally used to indicate the state of awareness that allows us to close our eyes and imagine, plan, predict, or estimate something. Consciousness allows us to solve problems by thinking rather than by relying on a trial-and-error approach. The debate about insect consciousness is complex because consciousness is difficult to demonstrate through experiments. Solid proof of consciousness in someone other than oneself, even if it's another human being, seems unattainable. It's even more difficult to consider achieving this with animals that can't tell us about their experiences. Even if they appear to behave like humans when placed in similar situations, we can never be sure that they'll have the same subjective experiences as us.

Despite the absence of direct evidence that insects have consciousness, they do show cognitive abilities that require what appears to be consciousness. In fact, if consciousness is the result of evolution, then it's likely to be of great use to a large number of animals that, like us, need to move around, explore the environment, recall things, predict the future, and cope with unexpected situations, even in a rudimentary form.

Several phenomena discussed in this chapter resemble manifestations of consciousness. Examples include the ability to recognize oneself as distinct from another entity, to plan, to recall specific events, and to take on the perspective of other individuals.[35] Western honeybees have an internal representation of space and time that we can observe through the symbolic communication of their waggle dance. This dance is sometimes even observed at night, suggesting that they retrieve their spatial memories out of context. Bumblebees appear to build mental images of objects, which allows them to recognize them in the dark. They also have an awareness of their own body size, allowing them to anticipate flight paths through obstacles and suggesting self-awareness. Western honeybees appear to be able to plan the construction of their nest's honeycomb cells within the physical constraints of the available space and tend to abandon a task when the risks of being wrong are too high, suggesting that they can gauge the consequences of their actions. Bumblebees show judgment biases associated with emotional states. Analysis of flies' brain activity shows distinct stages of sleep and wakefulness. At night, their brains behave like ours do during deep sleep and dreaming.[36] Biologist Andrew Barron at Macquarie University in Sydney and philosopher Colin Klein at the Australian National University in Canberra have proposed that insects' ability to know their position in the environment so as to orient themselves is sufficient evidence of subjective experience—one of the most basic aspects of consciousness.[37] According to their reasoning, this consciousness might even be located in a specific area of the brain: the central complex (the functional equivalent of our midbrain). The central complex is involved in navigation, and it processes information from the body and the external world. It likely allows insects to construct a representation of the world sufficient for subjective experience.

If we apply the same behavioral and cognitive criteria to insects as we do to vertebrates, then insects can be considered conscious, with no less certainty than we have about cats or dogs. Does every form of intelligence have to be accompanied by consciousness, however? With the development of artificial intelligence, the automation of cognitive tasks, such as facial recognition, conversation simulation, mathematical theorem solving, and artistic creation, has been made possible. Despite this, there is no reason to think that consciousness will appear in artificial-intelligence programs. If we could create robots capable of reproducing and evolving, and let them compete, it's not certain that conscious robots that have developed subjective experience, emotions, and the ability to predict the outcomes of their behaviors would prevail over robots without consciousness. In animals, consciousness has probably favored the emergence of all these cognitive abilities in order to cope with the constraints of a brain that is limited in the amount of information it can process and memorize. The most likely outcome is that we'll never be able to produce a conscious robot without discovering and replicating the neural basis of consciousness in our own brains.

4

The Superorganism

Sometimes research is like a video conference call: it should work, but it doesn't! Simon Klein, my first PhD student, learned this the hard way. He hypothesized that if insect intelligence is truly comparable to human intelligence, then we should be able to observe a high degree of variability in learning and memory abilities between insects of the same species. We all know people whose intellectual abilities are superior to ours, at least in some respects. One of my office colleagues, for example, is an incredible multitasker, whereas I'm more of a monotasker.* Like eye color and hand shape, cognitive abilities naturally vary between individuals. Psychologists observe this variability by measuring people's intelligence quotient (IQ). Evolutionary biologists see this variability as a source of adaptation that is at the root of humanity's success. Ethologists, meanwhile, wonder if the mechanisms that produce cognitive abilities are the same throughout the animal kingdom. As Simon asked, can we apply what we know about human intelligence to insects?

Simon had told me about a study by Gene Robinson and his team at the University of Illinois that had just shown that in western honeybee colonies only a handful of forager bees did more than 50% of the food collection, while a majority seemed to do little or

* The office gossips would say that my other colleague is only a half-tasker.

none.[1] Mainstream newspapers had reported this finding and suggested the existence of hyperactive and lazy foragers. But perhaps these bees simply had different flower-recognition and orientation skills?

We decided to test the existence of these "super-foragers" among bumblebees to understand their origin. Since it was winter, we had to find a room in which we could let insects fly. A former molecular-biology laboratory would do. Once emptied, we covered the windows to block daylight, installed high-frequency lights to prevent the insects from seeing in stroboscope and bumping into the walls,* added a heating system, built an airlock with mosquito netting to prevent the insects from escaping, and repainted the walls white to create the most consistent environment possible. Simon had also spent a lot of time building artificial flowers, each equipped with a scanner that read the RFID chips we used to tag our bees.† This would allow him to recognize the bumblebees every time one passed over a flower. Finally, Simon had repurposed large cardboard cubes, cylinders, and pyramids to serve as visual landmarks for the bees. The idea was to compare bumblebees' ability to create foraging routes in different environments presented in succession. Imagine that you're locked in a huge windowless hangar and asked to find the shortest path to collect five hidden objects. The visual cues in the room would be the only clues available to find your way around and learn the path. Therefore, if someone changes the location of these cues or introduces new ones when you're not looking, you'll think you've been teleported to a new environment, and

* The temporal resolution of a bee's eye is 100 hertz, which is about twice as high as ours. In theory, a bee can therefore see the succession of static images scrolling by as films projected at 24 frames per second and the flashing of standard 50 hertz neon lights, both of which are invisible to our eyes.

† RFID (radio frequency identification) chips contain information that can be read by dedicated scanners. This technology is frequently used to store personal data in transit cards, passports, and product barcodes. RFID is useful in ethology for tagging individual animals.

you'll try to learn a new path. At least that's what we wanted to make the bumblebees think.

The day before the experiment was to start, we installed three bumblebee colonies in our flight room, making sure that it was securely closed so that the insects wouldn't escape. We also set the temperature at 77°F (25°C) and sufficiently fed the colonies so they'd spend the night in the best possible conditions. We were finally about to find out if some bumblebees are more intelligent than others.

Despite all these precautions, the next morning Simon discovered a murder scene. Ants had invaded the laboratory, their columns connecting each of the bumblebee colonies and disappearing under one of the walls. All the bumblebees carefully microchipped the day before had been decapitated or de-antennated, and their stores of sugar water and pollen raided. I've already told you about the invasive Argentine ants that form supercolonies covering several thousand kilometers.* It turned out that our building was located on the territory of one of these colonies. In winter, when it's cold and food is scarce outdoors, the ants invade the building and tap into even the smallest of food sources by using their incredibly efficient system of pheromone trails. Scout ants are capable of recruiting thousands of their conspecifics to any source of sugar, protein, or li in just a few minutes. And that night, they had found our bumblebee colonies. After discovering the scene, I went to my office, two floors up, to find there had been an ant invasion there too. This time, the columns of marching ants connected the flowerpots on the windowsills and the shelves. The colony had found a water source. After this tragic incident, Simon decided to spend all his winters in Australia to avoid any more ant problems. So we postponed these experiments until the following summer, but I'll come back to that work later.

* See chapter 2, The Fragrance of Déjà Vu.

Colonies of social insects are often compared to "superorganisms."[2] This doesn't mean we should imagine ants with superpowers, a cape, and red underwear, but rather as a collective entity composed of individuals, just as our body is composed of cells. This superorganism has a super-skeleton (the nest), a super-thermoregulation system (through heat generation or air ventilation by the workers), super-energy reserves (honey), a super-reproductive system (the queens), a super-stomach (the larvae that demand food from the workers and digest it), and a super-immune system (the workers that clean the larvae).

To continue the metaphor, the colony also has a super-brain with a collective intelligence resulting from the coordinated action of all its members behaving as a single organism. This collective intelligence allows colonies to produce group behaviors that achieve results that far exceed those produced by an individual. Think of the architectural perfection of western honeybee nests, for example. Or the termite cathedral mounds constructed from mud that can reach more than 10 meters (about 33 feet) in height and inside of which is an extensive multilevel system of tunnels as well as an air-conditioning system, while each termite individually measures only a few millimeters. Even on a human scale, these are Herculean tasks! In South America, leafcutter ants build underground nests that can contain more than seven thousand chambers spread over an area of 150 square meters (nearly 500 square feet) and reach a depth of 8 meters (over 25 feet).[3] These nests are true megalopolises with millions of inhabitants, as well as highways, roads, gardens, and tunnels. The worker ants create huge marching columns to cut leaves from trees and store them in the nest. They use the leaves to cultivate fungus and then feed it to their larvae, a behavior considered a form of agriculture. These insects use chemical communication signals, which allow millions of brains to be coordinated in achieving a common goal. In the case of our invading Argentine ants, that common goal was to decimate our bumblebees. But how were the ants able to organize so quickly? Who gave the orders? Were they the decision of a single particularly influential ant in the colony? Or did the group hold a democratic vote?

Simple and Complex

A decade after Karl von Frisch* discovered the waggle dance, one of his students, Martin Lindauer (1918–2008), unraveled the mystery of collective decision-making in insects by documenting the swarming behavior of western honeybees. In late spring, when nests become overcrowded, new queens are produced, and bees swarm to establish a new colony—this is how the superorganism reproduces. About two-thirds of the workers in the mother colony leave with the old queen to find a new nesting site. The remaining third stay in the original nest with the new queen. The departing bees fly away and gather to form a cluster, often on a tree branch a few dozen yards from the mother nest. During this emigration process, the bees are vulnerable to predators, and the workers are preoccupied with protecting their queen in the swarm. They generally don't sting or disperse during swarming, allowing beekeepers to move swarms with their bare hands or even to let bees b safely land on them.

By carefully observing these swarms, Lindauer noticed that some bees were also doing a waggle dance and that this communication was key in deciding where to build the new nest.[4] Some dancers returned to the swarm without food and were sometimes covered in dirt. They weren't foragers but rather scouts in charge of locating potential nesting sites that could accommodate the colony. So how do the scouts inform the swarm about these potential sites? Hundreds of workers search for cavities and evaluate each one

* See chapter 1, A Poor Sense of Direction.

according to its size, its level of darkness, its isolation from the wind, and the protection it offers against predators. Then each scout dances to indicate the quality and location of the best site she has found. Contrary to the dances used during foraging, these dances are expressed on the surface of the swarm, directly referring to the position of the sun. The more a bee repeats a dance, the better the quality of the identified site. The dance therefore has an amplifying effect that stimulates the other scouts to go evaluate the site in question themselves and to dance in turn.

More recently, this research has been taken up by Tom Seeley,* a professor at Cornell University, an outstanding popular-science writer, and a mentor to an entire generation of researchers. I had the opportunity to attend one of his classes in conjunction with a conference. Someone asked him the question, "How do you become a great scientist?" He went to the board and simply wrote: "Do what you want to do." Then he left. Now that I'm an established scientist myself, I realize this is the best advice to give a student. I would only add: "And have fun!" Seeley and his associates at the University of Sheffield in the United Kingdom discovered that some of these scout bees can also have an inhibiting effect on dances.[5] In fact, bees that have visited different sites from the one signaled by a dancer may express a "stop signal" to dissuade the dancer from continuing its dance because the former are convinced that one of their sites is better. These bees emit a sound, produced by the vibrations of their wings, while striking their heads against the dancer they want to stop. The more the dancer receives these negative stimuli, the less she dances, eventually stopping completely. The stop signal's inhibitory effect is essential in order for a majority in favor

* Throughout his career, Tom Seeley has studied the collective decision-making of bees. He has published several best-selling popular-science titles, including *The Wisdom of the Hive* (1995), *Honeybee Democracy* (2010), and *The Lives of Bees* (2019).

of a single site to emerge within the colony. When a critical mass of scouts agree and therefore indicate the same site with their dance, a "consensus" is reached, and the voting stops immediately. The swarm flies to the chosen site and, in many cases, it proves to be the best site available to accommodate the colony. In this superorganism of bees, consensus thus emerges from the interactions among the scouts, without the intervention of a leader or a comprehensive plan to organize the vote. In fact, no single bee visits all the sites and has a complete view to make the best decision. It's the balance between amplifying the positive dance and the negative stop signal that allows for a collective, fast, and efficient selection of the optimal site.

Bafflingly, the collective decisions made by bee swarms abide by the universal laws of physics and chemistry that govern the behavior of atoms, molecules, organisms, societies, populations, communities, ecosystems, and—on a much larger scale—the universe. In all these "self-organized" systems, the elements interact locally according to simple rules, which can lead to the emergence of complex collective phenomena whose behavior is difficult to predict. This is what we call "complex systems." Much like a snowflake or a financial market, an insect colony is a complex system that operates with limited interactions between its elements (the interactions between two scout bees), positive feedback loops (a dance in favor of a particular nesting site), and negative feedback loops (the stop signal communicated to a dancer). After a certain threshold (critical mass of dancers), the system can change state or move from an unstable phase (the swarm is sitting on a branch in view of predators) to a stable phase (the bees all fly together to a single nesting site).

The Wisdom of the Crowd

While in mathematics $2+2=4$, this isn't always the case in insect societies. As I've said, these societies are complex. One of the incredible properties of these superorganisms is that their collective intelligence can be greater than the sum of their individuals' intelligence. This is called the "wisdom of the crowd."*

Jean-Louis Deneubourg and his fellow physicists at Vrije Universiteit Brussel in Belgium have conducted numerous experiments to document how this type of collective wisdom can emerge from simple individual behaviors in ant colonies.[6] Following in the footsteps of Ilya Prigogine (1917–2003), winner of the Nobel Prize in Chemistry for his work on self-organization in thermodynamics, Deneubourg and his team demonstrated how these universal principles can explain collective behavior throughout the animal kingdom. In their study of ant colonies, these pioneering researchers examined the astonishing phenomenon of ant trail optimization. They conducted the experiment by placing colonies of Argentine ants in plaster nests connected by a bridge to a source of sugar water. The bridge consisted of an arm that separated into two branches—a short one and a long one. Both branches rejoined the arm that led to the other end of the bridge. Even though the ants couldn't see the length of the branches before choosing one or the other, it didn't take more than a few minutes of the ants exploring the setup before most of them were using the shorter arm. Computational models developed to simulate ant colonies showed that the appearance of these behaviors could be explained by the use of a chemical marker (called a trail "pheromone") deposited by the ants while crossing the bridge. The trail pheromone acted as a feedback loop. Initially there was an equal probability of the ants using either branch, since the branches were unmarked. Then, as the insects used the bridge, the shorter branch became systematically more marked by the trail

* In 1906 at a country fair, the British mathematician Sir Francis Galton (1822–1911) asked about eight hundred people to estimate the dressed weight of a slaughtered ox. Galton discovered that the median guess was accurate within 1% of the weight measured by the country-fair judges. He called this phenomenon the "wisdom of the crowd."

pheromone because more ants had used it. This difference in chemical marking was sufficient to amplify the ants' attraction to the shorter branch and lead to a collective optimal choice for the colony without any individual ant being aware of having made a decision.

By the same process, ants are able to collectively choose the best available food sources in their environment.[7] In another experiment, Deneubourg and his colleagues presented colonies of black garden ants with sources of sugar water. In the presence of two identical sources, the ants selected one of the two at random. But when the two sources differed in sugar concentration, the colonies almost systematically developed a trail-recruitment system to the richer source. The scout ants deposited trail pheromones during their return journey to the nest and modulated the quantity of chemical markers according to the quality of the discovered resource. The more marked a trail is, the more likely ants are to follow it, find food, and reinforce the trail in turn, resulting in the selection of the best available resource.

Takao Sasaki and Steve Pratt at Arizona State University demonstrated how collective intelligence surpasses individual intelligence by comparing the behavior of ant colonies to that of single ants.[8] The two researchers leveraged the ability of *Temnothorax* ants to make collective decisions to find a new nest. These ants occupy small gaps under rocks. Their decision process looks similar to what western honeybees do during swarming, except that ants don't dance. They recruit other ants by forming tandem pairs in which the leader ant carries the follower ant on its back to the location of the discovered site. Once there, the leader lets the follower go so that she can judge the quality of the site for herself, which she does by assessing the site's brightness, the size of the hole, the diameter of the entrance, and many other parameters. The two ants then return to the nest to recruit others, and so on and so forth. Recruitment produces a positive feedback loop that leads to a consensus when a certain number of ants recruiting others to the same site is reached. Sasaki and Pratt were able to demonstrate that isolated ants had more difficulty choosing between four good-quality and four poor-quality nests than between only one good-quality and one poor-quality nest. The colonies, on the other hand, succeeded in selecting a good-quality nest under both conditions. When the ants were isolated, they had to each individually study all the solutions. In colonies, however, each ant visited fewer sites on average, but all sites were visited overall, increasing the likelihood of choosing a good site because of a shared cognitive load among scouts. Even so, this group advantage has its limitations under certain conditions. Sasaki, Pratt, and their colleagues showed these limitations by comparing the performance of colonies and individual *Temnothorax* ants in the perception task of choosing the better of two nest sites based on brightness level.[9] The darker the site, the higher its quality is assessed to be. Colonies were better able to select the best site overall when the illumination differences were small (both sites were dark). Individual ants, however, outperformed colonies when the illumination differences were large (a darker site and a brighter site). In the case where brightness discrimination was more difficult, a colony's positive feedback loop allowed it to more effectively assess small differences than a single ant could assess alone. When the discrimination task was easier, the experiment results suggest that the colony's positive feedback loop via pair recruitment could lead to overly rapid decision-making and lock the colony into a hasty and nonoptimal choice.

Contrary to what we may sometimes conclude from these experiments on bees and ants, there is no link between the complexity of social interactions and the complexity of collective behavior. Collective decision-making and swarm-wisdom phenomena can be observed in a large number of species, including nonsocial species like flies. Once, watching a colleague in Australia work feed her *Drosophila* (fruit flies) in culture bottles with two identical food sources on either side, I noticed that the flies tended to cluster on only one of the two sources instead of spreading out evenly to reduce competition. Despite the absence of recruitment by a dance or a pair, the fruit flies were indeed making collective choices! I then performed "cafeteria" experiments in which a fly or a group of flies had the choice between several food sources of different qualities. I made food patches that looked and smelled all the same but had a ratio of protein to sugar that varied by a factor of two (cooking for flies is pretty simple, but you have to be careful about the amounts of sugar, yeast, and dye). As I watched the flies move through this miniature self-service cafeteria, I observed that grouped individuals were much more efficient at finding their preferred food source hidden amid less-desired food sources than isolated individuals.[10] Flies in groups could simply join other flies that had already chosen a food source without second-guessing their decision. The better the quality of the food, the longer a fly would stay on it and tend to attract others. It's therefore possible to observe collective-intelligence phenomena comparable to those of bees and ants in other insects without recruitment or sophisticated social interactions.

It Takes All Kinds

The simplicity of the self-organization principles that govern collective behavior could give the impression that collective intelligence in insect societies emerges from interactions between identical individuals. All animal groups are characterized by some level of behavioral diversity, however, and a growing body of work suggests that this diversity is key in producing adapted group behavior. The organization of insect colonies is based on this diversity with a division of labor and a caste system of workers, each specialized in well-defined tasks. Even within a caste, individuals may show significant physiological, morphological, behavioral, and cognitive differences that will affect the functioning of the group. In western honeybees, for example, not all foragers are equally sensitive to sugar. Some can perceive very low concentrations in nectar, while others perceive only high concentrations. This variation explains the division of labor between foragers that have a low threshold of sensitivity and specialize in pollen collection and those that have a high threshold and specialize in nectar collection.[11] More generally, insects show stable behavioral differences over time that strongly resemble the way our personalities show distinct differences.

Despite his misadventures with ants, Simon Klein eventually carried out his "tele-transportation" experiments to study bumblebee navigation.[12] Over the course of two winters in the southern hemisphere and two summers in Toulouse, Simon observed bumblebees foraging in a field of artificial flowers that he regularly rearranged to make the bees believe they were in a new environment. The results can be summarized by the behavior of two bumblebees in particular. Bumblebee Y12* (let's call her Anna) was the one that excelled in the foraging task. After only a few minutes in a new environment, Anna was able to use the optimal route to reach each flower and return to the nest. This same bee was consistently the fastest to find the solution and the most consistent in using this optimal route in all environments tested. Conversely, bumblebee B03 (let's call her Wanda) was consistently the slowest to establish a route and the least consistent in following it over time. Anna and Wanda were the same age and size, had the same foraging experience, and came from the same colony. But they had different personality traits.

* The bumblebees were labelled with colored number tags so they could be individually recognized. Beekeepers use these tags to identify the queens in their bee colonies. The bumblebee I'm referring to here was tagged as yellow 12.

Lars Chittka and his colleagues in Germany and Australia demonstrated that behavioral differences in bees are related to differences in decision-making abilities.[13] To show this, the researchers tested bumblebees' ability to differentiate between color images of flowers projected on a wall. The point of using virtual flowers was to precisely control the colors presented to the bumblebees. Initially the bumblebees had to match one color (blue) with a sugar-water reward and another color (yellow) with the absence of a reward. All the bumblebees were able to learn this task with varying degrees of efficiency. Some made their decision quickly and were often wrong, while others were much slower to decide but rarely wrong. Bumblebees thus vary in the way they resolve the trade-off between speed and accuracy. Chittka and his colleagues then modified the consequences of making a mistake, punishing the bees with quinine (a bitter substance that bumblebees dislike) when they chose one of the two colors. In this new situation, the bumblebees were more efficient overall at solving the task. Individuals who were fast but not very accurate in the first task, however, were still fast but not very accurate in the second task; and the slow but accurate individuals were still slow but accurate. Once again, nothing distinguished the bumblebees from each other apart from their cognitive abilities.

It has been suggested that this natural variability in worker personality traits is an asset to the colony, allowing the insects to avoid putting all their eggs in one basket, so to speak. In fact, these personalities are all possible strategies for coping with new situations, such as a sudden change in environmental conditions. It's important for a colony to have individuals who focus on exploring and others who focus on exploiting known resources. James Burns at the French National Centre for Scientific Research (CNRS) in Gif-sur-Yvette, France, and Adrian Dyer at Monash University in Melbourne, Australia, have tested this hypothesis in honeybees.[14] They trained twelve foragers from the same colony to differentiate between two artificial flowers of different colors containing either sugar water or plain water. The experiment showed that foraging efficiency was higher for fast but inaccurate bees when the proportion of rewarded flowers was high and thus the risk of being wrong was low. Conversely, slow but precise bees had an advantage when the proportion of rewarded flowers was low and therefore the risk of being wrong was high.

In a colony, the proportion of individuals with different personalities can influence collective behaviors and define a personality of the whole group. Beekeepers are well aware of this phenomenon, given the existence of certain colonies that are more aggressive than others despite a similar genetic origin and breeding conditions. This is also the case with processionary caterpillars that form colonies in nests made of silk on tree branches. These caterpillars make collective decisions to search for food as a group, traveling in nose-to-tail processions by means of tactile and chemical signals. In these impressive larval colonies, some caterpillars are more active while others are much less so, and the proportion of both types of individuals influences collective decisions.[15] Therefore, colonies with a majority of active caterpillars are much less cohesive than those with a majority of inactive caterpillars. Lower cohesion is an advantage when caterpillars are in the presence of several nutritionally imbalanced but complementary food sources, as lower cohesion allows the caterpillars to leave the group to use the different food sources and regulate their diet. Conversely, caterpillars in inactive colonies concentrate on a single low-quality food source. The coexistence of these two strategies in natural populations is likely an evolutionary solution to maintain group cohesion while optimizing the caterpillars' ability to locate food sources and regulate their diet.

Insects and Robots

The principles of self-organization are so universal that we're able to program simple robots to reproduce the collective-intelligence phenomena observed in insects, however complex these phenomena may be. Michael Krieger, Laurent Keller, and their colleagues at the University of Lausanne in Switzerland had fun recreating the foraging behavior of ant colonies using swarms of robots.[16] In these experiments, the robots didn't look like ants at all but rather like electronic cards on wheels equipped with a battery, a motor, light sensors to find the nest, and a radio module to communicate with the other robots. This minimalist robotics approach is reminiscent of behavioral ethologists' approach in the early 20th century to describing animal behaviors by reducing the behaviors to simple responses to external stimuli.* Here the robots were programmed only to find plastic cylinders (representing food) in an enclosure and bring them back to the point of origin (representing the nest). They were aware of their colony's food requirements, but they didn't know the location of the feeding sites in the enclosure, and each trip cost them energy that had to be offset by food. Finally, they all had different probability thresholds of engaging in this foraging task (akin to personalities). The research team conducted a series of experiments with these robot colonies just as they would've done with ants. In particular, the researchers showed that the robots were able to self-organize to produce a collective foraging behavior that allowed the colony to meet its energy needs. The more robots there were in the colony, the more efficient this cooperation was, and implementing a recruiting system in which experienced robots were able to recruit a waiting robot (inspired by the recruitment method

* Some writers, like John Steinbeck (author of *Of Mice and Men*), have also been accused of being "behaviorists" at times because they tell stories without exploring their characters' behavior.

of tandem running observed in ants) significantly increased group efficiency. The experiments demonstrated that collective intelligence could emerge in groups of robots that were not programmed for that purpose.

Following these experiments, other researchers have gradually integrated autonomous robots into insect groups to manipulate their collective decision-making. Jean-Louis Deneubourg and his colleagues in Brussels succeeded in making groups of cockroaches choose a shelter that they would normally have rejected.[17] I was able to watch these experiments, and the result is mind-blowing. The original idea is similar to the storyline of Fritz Lang's film *Metropolis*, in which a scientist builds a human-looking robot that urges the workers to rebel against the city's master. Except that unlike in the film, Deneubourg is not a (completely) mad scientist. The Belgian team relied on cockroaches' natural tendency to make collective decisions when looking for a dark place to aggregate. The experiment's setup consisted of a large circular enclosure with two red plastic discs suspended 3 centimeters a little over an inch off the ground to serve as shelters. (Since cockroaches are not sensitive to the color red, they perceive these shelters as dark places.) The open-air enclosure was surrounded by an electric fence to prevent the cockroaches from escaping (which is never well received by colleagues in the laboratory). In this setup, groups of cockroaches made collective decisions according to a single feedback rule: the more cockroaches there were under a shelter, the longer they would stay under that shelter. Therefore, when the cockroaches had a choice between two identical shelters, they chose one at random. If they had a choice between a brighter and a darker shelter, they chose the darker one. Next, Deneubourg and his colleagues placed robots in the middle of the cockroach groups. The robots, still mounted on wheels, were perhaps five times as tall as any of the cockroaches. On the advice of Colette Rivault, my dissertation director at the University of Rennes, the robots were covered with the cockroaches' scent, so that they'd be perceived as being of the same species.* When the robots were programmed to have the same light preferences as the cockroaches, the mixed cockroach-robot group chose the darker shelter. But when the robots were programmed to prefer brightness, the robots disrupted the collective decision-making process and most often succeeded in inverting the cockroach groups' natural choice. Four robots were enough to persuade twelve cockroaches. As with any collective decision, the process was subject to a degree of uncertainty, and sometimes, but rarely, it was the cockroaches that influenced the robots.

Despite several attempts, there is still no autonomous robot capable of integrating into a group of bees. It is possible, however, to manipulate bees' collective behaviors with an automated mechanical arm. Tim Landgraf and his team at Freie Universität Berlin used this technology to build a waggle-dance simulator that could disrupt the information transfer during western honeybees' collective decision-making process.[18] The simulator worked by vibrating the plastic wing of a bee replica made from a piece of sponge covered in plastic wrap. The sound of the vibrations, imitating the wing movement of dancing bees, was amplified by a loudspeaker attached to a metal rod. The bee robot, named RoboBee, which had acquired the colony's scent prior to the experiment, could deliver sugar water to bees through a thin tube to mimic the exchange of food between the bees in the nest. With this device, Landgraf and his team demonstrated that bees could follow the simulator when it was programmed to dance and affixed to a wax frame in the hive. Therefore, bees trained to visit an artificial flower in sites A and B on either side of their hive could be led to visit new site C where there was no flower by following the dance indicated by RoboBee.

* See chapter 2, The Fragrance of Déjà Vu.

More recently, Franck Bonnet and his colleagues at the Swiss Federal Institute of Technology in Lausanne introduced robots into groups of honeybees and zebrafish,* facilitating the interaction of these two species.[19] The researchers achieved this feat by placing western honeybees in an enclosure containing two immobile robots producing varying levels of heat. Since bees are gregarious, their attraction to heat generated a collective decision to move to the hottest robot, either on the right or left side of the enclosure. In the second part of the experiment, Bonnet and his colleagues placed zebrafish in a circular aquarium containing robots that moved in the water among the fish. The fish robots influenced the swimming direction of the real fish to be clockwise or counterclockwise. The researchers then put the robots from the two sites—680 kilometers (423 miles) apart in Switzerland and Austria—in communication so that they could influence one another. The swimming direction of the fish robots determined which bee robot would produce heat and vice versa. That meant that if the school of fish turned to the right (clockwise), the movement influenced the fish robots, which increased the temperature of the bee robot on the right side of the enclosure and attracted the group of bees to its side. Correspondingly, if the bees chose the robot on the left, then the fish robots would turn left (counterclockwise) and influence the swimming direction of the school of fish. In this game of mutual influence, the researchers observed the emergence of consensus to go left or right in these two animal groups that have no reason to meet in nature and even less reason to communicate. By using the rules of self-organization found in insects, it is thus possible to integrate robots into animal groups and manipulate their collective behaviors. We can imagine that in the near future robots themselves

* The zebrafish (*Danio rerio*) is native to Asia. It's a popular aquarium fish and is also widely used in scientific research. It's the fish equivalent of the mouse for mammals and the fruit fly for insects.

will be able to learn from biological systems and evolve among animals to integrate into groups. Not so far off from *Metropolis*, I tell you . . .

This Election is Rigged!

Can we learn from all these collective behaviors? This question isn't entirely new. Aristotle and many other scientists, writers, philosophers, and politicians throughout history have pointed out that there is probably much to be learned from the social organization of insects. What is new, however, is how much we understand about these phenomena. The study of collective decision-making in bees, ants, flies, caterpillars, and cockroaches has helped to unravel the mystery of how collective intelligence might have emerged according to simple and universal rules. Beyond insects, these same general principles can be found in schools of fish, flocks of birds, herds of mammals, and hordes of primates. For thousands of years, natural selection has shaped these communities and their behavior to arrive at the best possible collective decisions. Now that we understand how these ingenious mechanisms work, we can use them to improve our own lives.

Engineers have long been interested in the subject of collective decision-making and have solved optimization problems using algorithms inspired by ant behavior,[20] such as the ability of colonies to find the shortest path through a maze. This form of artificial intelligence with a low cost in computing resources is widely used for improving transportation and communication networks. Other behaviors, like the division of labor among workers with different response thresholds for engaging in a task, have been used to optimize the task schedules for machines in factories. Ants' ability to transport their larvae and cluster them into orderly broods has also been applied to sorting information into complex datasets, detecting false information in social networks, and coordinating the actions of warehouse robots.

More recently, biologists and psychologists have also begun to draw inspiration from insects. After all, our societies are also complex systems, and we're part of superorganisms in which we're dependent on each other for shelter, food, and our physical well-being. Our interactions lead to the emergence of collective behaviors that are difficult to predict, such as elections, social norms, and even market dynamics. It's therefore quite relevant to use self-organization principles to improve how our societies function. We make group decisions all the time, whether choosing a restaurant or representatives to govern us. These decisions are fundamental for our well-being and the future of the planet, but unfortunately they're not flawless in terms of rationality, democracy, and collective interest. For example, in 2012 François Hollande defeated Nicolas Sarkozy in the French presidential election with 51.62% of the vote, but each of these candidates would have lost to François Bayrou had he been in the second round. In 2016 Donald Trump won the US presidential election against Hillary Clinton with votes influenced by often false and manipulative media sources.

To improve these collective decisions, Tom Seeley proposes we draw inspiration directly from bees.[21] During swarming, western honeybee colonies searching for a new nesting site show that three ingredients are necessary for simple, effective, and shared collective decisions. To make good decisions as a group, a set of possible options must be identified first. Bees do this through the coordinated action of several hundred scouts that visit different sites, perhaps determined by their different personalities. Multiplying efforts increases the likelihood that one of the identified options will be of high quality.

To reach a collective decision, information must next be shared among the population. If an individual doesn't make a discovery public, the information is unused, and this can lead to less effective group decisions. Bees use the waggle dance to indicate the location of potential nesting sites they've discovered. Each bee is free to judge the quality of a site independently, and the more often that site is judged to be good, the larger the number of scouts who visit it and recruit new bees in favor of that site, creating a social feedback loop.

Finally, for this collective decision to be rational, the information must be aggregated and the best solution chosen from among those identified. Bees do this by leading a frank debate between the scouts, who argue for different sites. This debate is somewhat like a political election, in that it involves multiple candidates (the potential nest sites), opposing ideas (the bees' dances), individuals loyal to certain candidates (the scouts who advocate for a particular site), and a set of undecided voters (the scouts who aren't yet advocating for a specific site). The outcome of the election is heavily biased in favor of the best site, because its advocates advertise it the most and convert others in favor of it faster. Finally, unanimous agreement is reached when all the scouts end up choosing a single site. In this collective process, each individual makes up its own mind through personal evaluation. Dissenting opinions are not suppressed. The debate is therefore open, and the final decision is based on the intrinsic superiority of the winning site, evaluated many times by several hundred individuals. A simple and effective model of democracy.

Our societies change extremely rapidly, however, and applying the principles observed in bee colonies on a much larger scale raises new questions.[22] In particular, over a short period of time humans have undergone a shift away from a system of small groups of hunter-gatherers solving local problems through vocalizations and gestures to a system of global decisions affecting all of humankind in order to endure pandemics, climate change, ecosystem degradation, the rise of extremism, the risk of nuclear war, and economic inequality. This change of scale is accompanied by faster and unfiltered new modes of communication in the form of huge dispersed networks connected by digital technologies. We don't know exactly how these changes impact collective decisions, but research on social insects already shows that group size (one group or billions of subscribers on networks), information flow (instantaneous or delayed), information quality (accurate or unverified), and the information acquisition system (direct or filtered by algorithms) are extremely important factors in the dynamics and effectiveness of collective behavior. All these elements, which are essential for developing our collective responses to global challenges, are still poorly understood and not easy to predict. Formulating a response to a large-scale issue amid these uncertainties would be like asking you to predict whether a video conference call with your coworkers will work or not . . .

5

Achilles' Tarsus

Thetis was a goddess. She was also an overprotective mother who found her marriage to a mere mortal very difficult to accept. Because their children were not immortal, Thetis was so concerned about their likely deaths that she sacrificed her first six children at birth. But at her husband's insistence, she spared their last son. To be sure that no harm would ever befall him, Thetis immersed him in the river Styx, whose waters had miraculous powers that could make someone invulnerable. Since the baby couldn't swim, she held him by one foot as she dunked him in the water. The child would now be invincible.

Thetis entrusted her son's education to Chiron, an erudite centaur. With Chiron as his teacher, the young boy had a wonderful childhood, learning music, literature, science, and how to handle weapons. Unfortunately, war broke out because of a love affair and the two camps on either side of the sea raced to assemble the strongest possible army to win. A soothsayer had told the Greek camp that somewhere there was an invulnerable young man and that this warrior would be the key to winning the war. This soothsayer was the same one who had warned Thetis a few years before of the inevitable death of her son if he ever went to fight in battle. When the war broke out, the goddess was afraid for her son and did everything she could to prevent him from joining. She even went so far as to disguise him as a girl. But Odysseus, a cunning war captain, eventually found the young man and enlisted him in the Greek army, along with his best friend, Patroclus.

As expected, the young man was formidable in battle. But after ten years of loyal service, he decided to retire from the army because of a disagreement with his superiors. It was then that his friend Patroclus, wearing his armor, was killed in battle by the greatest of the enemy's warriors: Hector.

When the retired warrior learned of this, he decided to avenge his friend. Wearing new armor, he slaughtered everyone in his path. Corpses littered the battlefields. Nothing could stop him. He eventually found Hector and killed him. This man was undoubtedly the greatest warrior of all time, and he ultimately led the Greeks to a final victory.

But as he led the assault on the enemy's capital, he was struck by a deadly arrow to his heel—his right heel, to be exact. The prophecy was fulfilled. The heel that Thetis had held and not dipped in the river Styx was his only weak point. And this fact had not escaped the god Apollo, who had seized the opportunity to take revenge for the death of his own son, Hector.

Insects share many similarities with our hero. Let's hope their story has a better ending though! Like Achilles, they're the heroes of many popular stories, and they dominate the earth. In fact, the oldest fossils of insects date back four hundred million years,[1] well before the appearance of dinosaurs. Insects are therefore among the first animals to have adapted to life on land. Today they're everywhere—on all continents and in almost all climates—successfully inhabiting both terrestrial and aquatic environments. Only the oceans remain free of insects. With nearly 1.3 million identified species, insects represent 85% of animals and 55% of global biodiversity. It's estimated that there are ten quintillion insects on the planet, a biomass three hundred times greater than that of all humanity, and four times greater than that of all vertebrate animals. All the ants alone would weigh as much as all humankind.[2] Insects are also essential to the functioning of our ecosystems. These small animals, often herbivorous, are eaten by many other larger animals. They're at the base of most food chains and perform a number of ecological functions essential to the maintenance of biodiversity, such as plant reproduction (through pollination and dispersal), recycling of organic matter, and predation, without which the world as we know it wouldn't exist.

Also like Achilles, however, insects are not indestructible. It seems that they're even more fragile than we might have believed. Today it's estimated that over 40% of insect species are in decline, and a third are in danger of extinction.[3] Bees are particularly affected by this biodiversity crisis, with entire geographic areas devoid of bees. You may have heard of the Chinese farmers in the mountainous region of Sichuan who have been pollinating crops by hand for decades owing to a lack of bees. In the last twenty years, European and American beekeepers have seen the annual loss of their honeybee colonies rise from 10% to 40%, and even 90% in some areas. The hives are emptying. We call this "colony collapse disorder." Jonathan Chase and his team at the Martin Luther University Halle-Wittenberg in Germany analyzed data from forty-one countries that clearly showed the global insect biomass has been declining 1–3% every year since the 1920s.[4] At this rate, all insects could disappear within a century, endangering the ecological services they provide and all biodiversity.

What would a world without insects look like? It's difficult to answer this precisely. What is certain though is that all terrestrial ecosystems as we know them would be greatly disrupted, and most of our food sources would disappear. Researchers in the fields of biology, ecology, and evolution are trying to understand the reasons for insects' decline so they can stop it. The study of intelligence in bees and some other insects discussed in previous chapters does show that these animals have a weak spot. This Achilles' heel, or rather "tarsus" because insects do not have a heel,* is their brain.

Every day we use large amounts of phytosanitary products to protect crops, release industrial pollutants into nature to produce objects and energy, destroy natural habitats to build houses and factories, and introduce new pests and diseases into geographic areas that were previously free of them. Some insects are more vulnerable to these combinations of stress factors. For those, like bees, that have developed sophisticated cognitive abilities in a compact, miniature brain, every neuron counts. The smallest attack on the nervous system by a synthetic molecule or a pathogen can lead to deformity or dysfunction. This can compromise a bee's normal behavior, affecting its ability to learn, forage, interact with fellow bees, and feed its young.[5] Here's why that's a problem.

The Harmful Effects of Nicotine

The widespread use of pesticides is often singled out as the main cause of insect decline. Science confirms this. Pesticides are substances used to get rid of crop-damaging organisms, such as aphids and caterpillars, which are considered "pests." When the target is insects, the type of pesticides used are called "insecticides." The discovery of a new class of molecules called "neonicotinoids" was a major turning point in chemical control of insect pests. The first neonicotinoid-based insecticide was marketed in 1991 under the name Gaucho, in reference to the skilled horsemen who protected the land and horse and cattle herds of South America. These "crop protector" molecules act directly on an insect's brain. Neonicotinoids

* An insect's leg is composed of several segments, which in order from most proximal to most distal are the coxa, trochanter, femur, tibia, and tarsus.

and nicotine—the latter long known as effective against moths and aphids—share similar chemical properties. Like nicotine, neonicotinoids bind to nicotinic acetylcholine receptors* located in an insect's central nervous system and block the function of affected neurons. In high doses, neonicotinoids cause paralysis and death.

Today this class of insecticides is the most commonly used throughout the world to protect crops and livestock. In 2015 neonicotinoids accounted for a quarter of the total global insecticide sales volume. There are about ten neonicotinoids, such as imidacloprid, clothianidin, and thiamethoxam (remembering these names could give you an advantage playing Scrabble). In modern grain farming, these insecticides are not simply sprayed on the crop; instead, the seeds are most often treated with one of these insecticides, which spreads through the plant tissue as it grows. This is the case for seeds of mass-flowering crops such as rapeseed, maize, and sunflowers. Foraging bees are therefore exposed to these neurotoxins contained in the nectar and pollen of contaminated plants, and in turn expose other members of the colony through physical contact or through the food they store in the nest. Insects are also exposed to huge clouds of dust containing these pesticides during the harvest when crops are ground up for harvesting, which sometimes also contaminates nearby water sources. Apprised of these poisoning risks, the manufacturers of these pesticides now ensure that the doses to which bees are exposed are low enough not to kill them.

* Neurotransmitter receptors are proteins embedded in the neuronal membrane at the junction between two neurons (called a "synapse"). They become activated when a neurotransmitter binds to them, and the effect of their activation varies depending on the neurotransmitter. In the case of nicotinic acetylcholine receptors, they're activated by the neurotransmitter acetylcholine, which is the chemical that activates muscles. This chemical reaction is terminated when the enzyme acetylcholinesterase breaks down acetylcholine. When neonicotinoids bind to nicotinic acetylcholine receptors, however, the activation cannot be terminated by acetylcholinesterase because this enzyme cannot break down neonicotinoids. This reaction overstimulates and blocks the receptors, ultimately causing paralysis and death.

While true that these reduced exposure levels don't cause death, even low doses of neonicotinoids can cause considerable brain damage in insects, making them no longer capable of solving the cognitive problems necessary to orient themselves correctly, recognize which flowers to forage, collect food, pass food on to others, and communicate. This damage snowballs, meaning that the whole colony is no longer functioning and is in danger of collapsing.

Mickaël Henry, Axel Decourtye, and their colleagues at the bee research unit at INRAE* and ITSAP Bee Institute in Avignon, France, were among the first to sound the alarm by demonstrating these effects on bees.[6] In 2012 they analyzed the ability of honeybees exposed to thiamethoxam (eighty-seven Scrabble points on a triple word-score square!) to orient themselves to return to their hive. To simulate poisoning, the researchers captured and then fed foragers sugar water containing a low dose of this pesticide. They also took this opportunity to equip each bee with a radio microchip (RFID) to be able to recognize them individually. Henry and his team then released the microchipped bees within a 1-kilometer (just over a half-mile) radius of their original hive, either in a site the bees were already familiar with or in a new site. Two days later, chip readers placed at the hive entrance gave the verdict: regardless of which site the bees were released at, a much lower proportion of bees exposed to the pesticide returned to the hive compared to unexposed bees. Their sense of direction had been severely impaired by the low dose of pesticide they had ingested.

The year following the publication of these results, the European Union temporarily banned using these compounds on grain crops. The scientific community has also undertaken numerous studies to understand the effects of these pesticides on other bee species. Ben

* INRAE is France's National Research Institute for Agriculture, Food, and the Environment.

de Bivort and his colleagues at Harvard University, for example, highlighted the general influence of neonicotinoids on bumblebee behavior.[7] Using video tracking of each bumblebee's activity in the nest (this time with QR codes printed on plastic tags and glued to the thorax), the research team was able to observe that exposure to imidacloprid reduced the social interactions and care provided by the workers to the larvae. These behavioral changes prevented the bumblebees, which normally contract their muscles to emit heat or fan hot air with their wings to cool, from maintaining a stable temperature in the nest. This inability to thermoregulate the nest at 30°C (86°F) seriously compromised larval development and colony survival. Geri Wright and her team at Newcastle University in the United Kingdom demonstrated an even more worrisome effect of exposure to these compounds: neither western honeybees nor buff-tailed bumblebees were able to avoid feeding on contaminated nectar when given the choice.[8] Quite the opposite, the foragers preferred the sugar solutions containing neonicotinoids to the uncontaminated sugar water—a preference which causes their premature death. This behavior, which can be described as "irrational," can be explained thanks to an observation made in a natural environment: bees also prefer nectar containing low doses of nicotine produced by certain plants to defend themselves against herbivores. Therefore, neonicotinoids, like nicotine, act like a drug that keeps bees coming back for more, a side effect of pesticide use that is disturbingly similar to cigarette addiction.

These studies provide a damning assessment of the effects of neonicotinoid-based insecticides. Thanks to the commitment of researchers and beekeepers, these studies have also resulted in much more restrictive regulations in the European Union, with a ban on the use of the three most common neonicotinoids on all outdoor crops beginning in 2018 and a fourth (thiacloprid) in 2020. Unfortunately, many European Union member states have requested exemptions so they can temporarily use these banned products. Moreover, as soon as these substances were banned, they were replaced by pesticides that are equally dangerous, such as sulfoximines (seventy-two points!). This relatively new class of insecticides is used in many industrialized countries, such as China, Canada, and Australia. Sulfoximines are attractive because they're effective against insects that have developed resistance to neonicotinoids, which means the insects are no longer sensitive to them. But the reality that these pesticides are applied by coating seeds, act on nicotine receptors in the brain, are nonspecific to the targeted species, and are insoluble in water makes them similar compounds to neonicotinoids, with devastating effects on pollinators. Mark Brown, Elli Leadbeater, and their team at Royal Holloway, University of London in the United Kingdom have shown that regular exposure of bumblebee colonies to low doses of this new sulfoximine-based pesticide significantly weakens colonies, causing foragers to collect less food, queens to lay fewer eggs, and larvae to take longer to develop.[9]

So everything is far from being resolved. In addition to scientists, beekeeping and nature-protection associations continue to warn the general public and decision-makers about the harmful effects of these substances on bees and insects in general. In fact, all insects, regardless of size, have nicotine receptors in their brain and are therefore targets of these pesticides. For example, beekeepers in the Occitania region of France have organized an event called "l'Abeille du 18 juin" (the bee appeal of 18 June)* to encourage the government to declare a state of emergency for agriculture and biodiversity. But the economic stakes are such that banning the use of insecticides entirely is a complicated decision in the short term.

* Referencing Charles de Gaulle's Appeal of 18 June 1940 calling for French resistance during World War II, this appeal on behalf of bees and biodiversity was issued in 2019: https://www.apiculteurs-occitanie.fr/communique-labeille-du-18-juin-2019/.

Alternative solutions exist, however, such as helping farmers adopt new agricultural practices that ensure profitability, like agroecology. A more judicious use of pesticides can also be accomplished by only exposing targeted species through the use of "Trojan horses," meaning bait traps introduced into an animal's nest. Pheromone traps to attract insect pests and the use of beneficial species to manage them can replace most insecticides. Better management of agricultural areas to conserve habitats and the use of crop varieties that provide high nutritional quality nectar and pollen can attract pollinators to barren areas and greatly increase agricultural production. All these approaches are avenues that research is exploring to provide viable and sustainable solutions for agriculture.

Rock Isn't Dead*

The other problem with neonicotinoids is that they get so much attention from researchers and the public that other sources of pollution are overlooked. Bees live in an environment that has been damaged by our industrial activities. These pollutants are also stress factors for insects, but unfortunately they're not studied, or at best infrequently. Recently I became interested in heavy metal pollution, inspired by a student determined to tackle this topic in her work on her PhD. Growing up on a farm, Coline Monchanin saw the deterioration of the bees' circumstances firsthand. She became acutely aware of their decline and the consequences this may have for biodiversity.

Before the matter of heavy metal pollution was brought to my attention, I must admit that the term "heavy metals" didn't mean much to me, aside from a vague memory of the periodic table† from

* This section discusses the problem of heavy metal pollution from rocks.

† The periodic table of elements presents all the chemical elements ordered by increasing atomic number and organized according to their electron configuration.

my school days and some legendary Led Zeppelin* guitar riffs. I quickly realized that heavy metal pollution was a major public health concern, on a global scale, and still largely underestimated. It's probably one of the most widespread forms of pollution on Earth for which we are directly responsible and on which we must act.

Heavy metals are constituent elements of rocks. They're naturally present in low concentrations on the earth's surface. During the 20th century, however, the intensification of mining, metallurgy, fossil-fuel combustion, industrial production, and agricultural use of these metals doubled their concentration in the environment. In low doses, some of these compounds, such as cobalt, copper, iron, manganese, and zinc, are essential to our diet and the diets of many other living things. But other metals, such as lead, arsenic, cadmium, mercury, and nickel, are extremely toxic. You might say the song remains the same!† We turn a blind eye while releasing pollutants into nature, only realizing much later that we're poisoning everything around us. In the case of toxic metals, these compounds contaminate the air, soil, water, and plants. In the 1970s, however, we became more aware and made efforts to reduce lead emissions to address lead poisoning, which causes behavioral disorders and intellectual disability in humans. Starting in the 1920s, lead was added to gasoline for motor vehicles to lubricate the engine and increase its performance, resulting in massive releases of lead into the atmosphere. In response to this public health risk, leaded gasoline has gradually been banned worldwide.‡ This is a first step. But it's not nearly enough.

* This English band from the 1970s is often considered one of the pioneers of heavy metal.

† *The Song Remains the Same* is a concert film by Led Zeppelin released in 1976.

‡ Leaded gasoline was banned in the United States in 1975 and in Europe in the 2000s. Lead is still used in aircraft fuel, batteries for motor vehicles, paint, jewelry, toys, and cosmetics.

To take stock of the state of affairs, Coline analyzed all the scientific articles since the 1970s that were concerned with the effect of heavy metal pollution on invertebrates.[10] The major lesson she learned was that most studies show clear negative effects on insect survival, physiology, and behavior, even at doses below the World Health Organization's recommendations. These dose recommendations, when they exist,* have been determined based on toxicity studies conducted in humans and laboratory mice under controlled exposures. It seems that insects are more sensitive, or simply more exposed, to heavy metal pollution than we are. Moreover, most studies have been conducted on pest or model species but almost never on insects that play ecological roles essential to the proper functioning of ecosystems, such as pollinators. And very few studies have been carried out on joint exposures to different metals, which is most often the case in contamination in nature. In fact, heavy metals are generally not present in pure form but are part of the composition of rocks or sediments in much more complex chemical forms. Coline's conclusions therefore show a significant gap between the research carried out in the laboratory and the realities in the field. Her conclusions call for an in-depth reevaluation of the real consequences of heavy metal pollution and of the permissible limits recommended by public health authorities, which must quickly be brought up to date.

Following this grim observation, Coline and I decided to carry out experiments to study the influence of these heavy metal pollutants on bee intelligence. Reluctantly, Coline fed western honeybees a mixture of sugar water and low doses of lead, zinc, and arsenic in order to evaluate the influence on their learning and memory capacities. In all these experiments, the compounds administered

* At the time of writing, there are international guidelines for only four heavy metals: lead, mercury, arsenic, and cadmium.

alone or as part of a mixture significantly reduced the bees' cognitive abilities, and they were no longer able to learn certain simple exercises, such as discriminating between two scents.

To evaluate the relevance of these laboratory observations, Coline then conducted a field study. She placed hives within a 10-kilometer (6.2-mile) radius of a former mining site in the south of France. To do so, she benefited from several kindhearted beekeepers who agreed to lend her their apiaries and their bees so we could carry out the study. This site, in the middle of the Black Mountain range, is a former gold mine that operated until the early 2000s. In addition to gold, miners extracted arsenic, which is still abundant today and is a worrisome source of water and soil pollution in the area. Over the course of several weeks, Coline visited the mine and tested the ability of bees at different sites within it to learn and form olfactory memories. The results were unquestionable: the bees that collected on the former mining site had markedly degraded cognitive abilities compared to those located 10 kilometers (6.2 miles) to the north or the east. Most of the bees at the mine were no longer learning. On the rare occasions when they did learn, their memory failed them. Today it's still unclear how heavy metals affect a bee's brain. But their impact is undeniable. Even in trace amounts, these metals are a source of harmful pollution for insects and probably for many more animal species that haven't yet been studied. Unfortunately heavy metals are still rarely studied and for the most part aren't regulated, except for those identified as toxic to humans.

The Badger and Preventive Measures

If only pollution were the only problem. Bees must also learn to live with an increasing number of predators and diseases that we spread across the surface of the globe aboard trucks, ships, and planes. These additional sources of stress on insects can affect their nervous system. Some parasites even manipulate the behavior of their host to facilitate their own reproduction and propagation. One such parasite is the Gordian worm, whose juveniles induce the suicide of their insect host by manipulating the host's water-seeking behavior. Once the host enters water, the worms can emerge and reproduce.[11] These nematomorphs then aggregate to form a mass during mating that resembles a knot. Their name derived from the legendary Gordian knot of King Gordias of Phrygia—a knot that Alexander the Great cut with his sword to become ruler of all of Asia.

Bees are exposed to a large number of parasites, including bacteria, fungi, mites, and other insects. These organisms are often present in the nectar or pollen of plants and can affect insects' cognitive abilities. The most famous of these is a small single-celled fungus known as *Nosema*. This parasite enters a bee's intestine in the form of spores, where it settles and uses the bee's glucose metabolism to reproduce. This invasion of the intestinal tract triggers nosemosis, a disease familiar to beekeepers that manifests itself in the form of diarrhea dysentary, learning disruptions, and altered foraging behavior. In the western honeybee, *Nosema* also disrupts the division of labor by accelerating the transition of workers from performing domestic tasks in the nest to foraging tasks outside the nest, so that younger and younger bees start foraging for nectar and pollen without being properly prepared for it. These young foragers with immature brains are much less efficient and have a greater risk of getting lost in flight than normal-aged foragers, which significantly weakens colonies.[12] Since its discovery in the 1990s in apiaries, forms of this parasite have been identified in many species of wild pollinators, including bumblebees, wasps, and butterflies, demonstrating the ability of *Nosema* to disperse and find new hosts. Fortunately bees are not completely passive. Some have developed self-medication mechanisms in response to these infections. Thus *Nosema*-infected workers tend to increase their intake of protein-rich pollen or propolis,[13] which boosts their immune defenses and their ability to cope with low levels of infection.

Sometimes we move bees to the wrong place and unintentionally put them in contact with their predators. A few years ago, we introduced about thirty beehives into the woods next to my laboratory in Toulouse. This apiary is a peaceful haven that contrasts with the university's bustling atmosphere created by its thirty-five thousand students and ten thousand employees—about the same population of a small city. I had a 200-square-meter (roughly 2,000-square-foot) tent set up there to study bees' navigation behaviors. I'd put a lot of thought into choosing this tent, which was actually a very long tunnel. It had a particular screen that allowed the necessary ultraviolet light to pass through, so that the bees would know the position of the sun in the sky, while also ensuring that they wouldn't leave the study area and that unwanted insects wouldn't interfere with the experiments. The day after the tent was installed, I discovered holes in the canvas a few centimeters off the ground. My excitement about the new equipment was quickly replaced by annoyance. My first thought was that the holes had been caused by cats, and I asked the students to stop feeding the felines on campus near the apiary. I also sprinkled coffee grounds, which are known to be a repellent, all around the tunnel. But nothing helped. Night after night, the holes multiplied and grew larger. When they reached the size of a soccer ball, I installed an electric fence around the tunnel. The problem only got worse. Then things progressed, and hives full of bees were knocked down at night. Some mornings, honey frames lay on the road outside the apiary. Then one day we discovered clumps of black and white fur. The criminals had left their signature. It turned out, a family of badgers living nearby had taken up the habit of tearing holes in my tent to feed on our bees and their honey. But knowing your enemy allows you to better coexist. So we redesigned the whole structure, positioning the hives higher and changing the layout of the apiary so that the tunnel was no longer in the badgers' path.

On occasion bees use control strategies acquired during their long coevolution with predators. This happens when they come across hornets on their route. Hornets are insect predators that are wild about bees, which are abundant sources of protein. These large social wasps capture bees as they're returning to the nest, when they're full of nectar and they fly slower. A bee's death is as fast as it is violent—the hornet decapitates the bee, taking only its thorax, which contains its wing muscles and thus a high level of protein. Hornets sometimes hover in large numbers at the entrance to a colony, limiting the foraging activity of bees, who don't want to confront these predators. But some bees have found a solution. Japanese honeybees,* for example, have developed collective defense mechanisms to protect themselves from predation by Asian giant hornets (murder hornets)†—including literally cooking them![14] When a hornet approaches, the bees lure it into the nest. They wait for their predator to start attacking, then they pounce on it. Each bee surrounding the hornet begins to vibrate, which increases the nest's temperature. The air around the hornet can reach 47°C (116°F). The bees emerge unscathed because they can withstand heat two degrees higher than a hornet can. But stuck under the mass of bees, the hornet ends up roasted alive.

These forms of collective defense are also observed against parasites and pathogens. In honeybees, workers show hygienic behaviors. When they detect cells containing diseased brood (eggs and

* The Japanese honeybee (*Apis cerana japonica*) is a domesticated bee native to Japan. It's a subspecies of the Asian honeybee (*Apis cerana*).

† The Asian giant hornet (*Vespa mandarinia*) is the largest known social insect species. It can grow up to 5 centimeters (nearly 2 inches) long. This hornet is native to Asia and was introduced to North America in 2019.

larvae), they destroy the larvae and clean the cells, thus limiting the spread of disease. These abilities constitute a form of social immunity whose effectiveness varies greatly depending on the bee population and species. Sylvia Cremer's team in Vienna, Austria, and Laurent Keller's in Lausanne, Switzerland, identified mechanisms of social immunity through avoidance, an even simpler mechanism and probably more prevalent in social insects, including bees.[15] In studying ant colonies' response to infection by a fungal pathogen, the researchers observed clear differences in the social-contact networks* of ants in healthy nests and those in infected nests. With graphical representations, in which the ants are nodes and their contacts are links, it became clear that the ants reduced their interactions in the presence of the fungus. Therefore, after foragers were exposed to the fungus, they tended to isolate themselves by spending more time outside the colony, which drastically reduced contact with uninfected colony members. Nurse ants, meanwhile, increased their brood-care activity and carried the brood deeper into the nest to reduce contact with infected individuals. This rapid response by both infected and uninfected ants helped to control the spread of the pathogen within the colony. It appears that insects have been taking preventive measures since the dawn of time, long before such measures became commonplace in our societies. This social-distancing strategy is a simple and efficient defense mechanism that potentially enables collective resilience to exposure of different stressors, such as pathogens, as well as pesticides and pollutants.

---

* The use of network analysis has revolutionized many areas of biology, including ethology in the 2000s. These tools allow complex social interactions to be visualized graphically and then reduced with simple mathematical descriptors. The scope of application is immense. For example, after having experimented with this type of analysis for a long time in bees, a colleague of mine in Toulouse began using network analysis to develop new soccer and rugby strategies.

The Second Brain

Doctors often say that the gut is a human's "second brain." Our digestive system has nearly two hundred million neurons. It's also home to many microorganisms that help us digest by breaking down and absorbing food. This intestinal flora, called the "microbiota," is made up of several hundred billion nonpathogenic bacteria. This corresponds to about 1 kilogram (over 2 pounds) of microorganisms, or approximately the weight of our brain. Our body is therefore a complete ecosystem in which the majority of cells that comprise it are not our own. This may seem frightening at first glance. But, as with all ecosystems, these interactions are the result of thousands of years of coevolution, and they create a subtle balance from which all members benefit. Numerous studies are beginning to show that these microorganisms aren't just passive beings that feed on our metabolism. They also communicate with our brain, influencing our immunity, our physiology, and some of our behaviors.[16] This research on humans and mice has recently expanded into the insect world. In particular, the research highlights the disruption of symbiotic relationships between insects and their microbiota owing to human influence on the environment, such as the use of antibiotic treatments for crops and livestock.

Most insects host much simpler bacterial communities than we do. For example, only about ten species are commonly found in the digestive tract of western honeybees and *Drosophila* flies. But there are notable exceptions, such as certain termites that feed on wood or soil, which are poor in nutrients, and whose intestinal flora is much more complex. These termites require a wide variety of symbionts to make these raw materials digestible or to provide the essential nutrients not available in the food, either because these symbionts produce useful metabolites or because they're ingested directly by their host. These bacteria are usually acquired from food.

In bees, ants, and cockroaches, bacteria are shared during social interactions and contribute to the establishment of the colonial odor. Yehuda Ben-Shahar and his colleagues at Washington University in St. Louis, Missouri, for example, demonstrated that when sister western honeybees were inoculated with different bacterial strains, they developed distinct chemical signatures and were no longer recognized as members of the same colony.[17] Microbiota can therefore have a significant influence on the behavior of insects and shape their entire social organization.

I had the opportunity to contribute to this body of research when I was studying flies in Sydney Australia. My daily routine was to have my morning coffee at a sunny table outside the laboratory. One day, my table was occupied by Adam Chun-Nin Wong, a postdoc fresh out of Cornell University who had the same coffee-in-a-sunny-spot idea as me. Adam was a specialist in fly microbiota. At that first meeting, he persuaded me to read a study conducted by Eugene Rosenberg's team at Tel Aviv University in Israel.[18] These researchers fed *Drosophila* flies a diet that was either rich in molasses (a simple sugar) or rich in starch (a complex sugar). Flies are usually bred in small plastic bottles. In the bottom of the bottle, food is placed that is solid enough for the adults to move around on without sinking in but also soft enough for them to eat, lay their eggs in, and for the larvae (maggots) to burrow tunnels into. A proper *Drosophila* farm contains thousands of these bottles and smells strongly of baker's yeast (the main ingredient of fly food). In this study, these different diets of simple and complex sugars were meant to promote the establishment of different communities of bacteria in the flies' intestines. When the flies from the bottles with different diets were mixed, they showed strange sexual preferences: the flies raised on molasses preferred to mate with each other, as did those raised on starch. This preference for sexual partners fed the same diet appeared in only one generation and disappeared just as quickly following antibiotic treatment. The researchers had just made a major discovery: bacteria in the digestive tract—selected by the flies' diet—influence their host's behavior, probably by modifying chemical communication between flies.

After a few more coffee breaks with Adam spent discussing these experiments, we agreed to test the influence of flies' microbiota on their ability to choose their food.[19] Within just a few days, Adam isolated strains of bacteria from my fly farm. He then created groups of flies without microbes (axenic flies), groups of flies with several strains of microbes (wild flies), and groups of flies inoculated with a single selected strain of microbes (gnotobiotic flies). We deprived the flies of food, then placed them in small boxes containing sugar-, yeast-, and agar-based diets that I had prepared. By observing these flies for several hours, we discovered that they were attracted to different food sources. When the flies were given a choice between several identical diets that were colonized by different populations of bacteria (much like choosing between variations on the same recipe for moldy cake), they preferred to eat the food colonized by the bacteria already present in their gut. When the diets offered were of different qualities (this time they had to choose between several different recipes for moldy cake), the flies seemed to weigh their attraction to the microbes they carried in their gut and their need for specific nutrients. In most cases, microbes seemed to win out over nutrients; so the microbiota sometimes led them to make irrational nutritional decisions. Imagine not eating for forty-eight hours and then your stomach makes you choose a moldy slice of cake over a well-balanced three-course meal.

Other research indicates effects of microbiota on cognition that are even more immediate, showing that loss of gut microbes reduces learning and memory performance in flies.[20] It's likely that microbes produce chemical compounds that communicate directly with the insect brain, as has been shown in humans where symbiotic bacteria are responsible for modulating many neurodegenerative diseases, emotional states, and social interactions. A key objective to protect biodiversity is to identify these compounds and produce probiotic supplements that would increase insects' cognitive abilities or that would promote the expression of key behaviors for conservation, such as plant pollination by bees.

Five Fruits and Vegetables a Day

Poor nutrition is a problem for bees. We're told day in and day out that a varied diet is essential for good health, and we need to eat five servings of fruits and vegetables a day, but nothing too fatty, nothing too sweet, and nothing too salty. Insects also have a vital need to regulate their diet and not eat just anything. For example, from a wide range of options, migratory locusts are able to select the plants that meet their carbohydrate, protein, lipid, and mineral needs.[21] No single plant allows them to achieve a perfectly balanced diet, so the locusts must taste the different plants available and ingest them in adequate quantities to balance their diet. Their dietary needs change over the course of a lifetime, and they adjust their choices accordingly. This type of change over time is also found in humans, whose diet varies considerably according to age.

In bees, the task of regulating diet is carried out collectively by the foragers, who have the important responsibility of feeding all members of the colony, which can be a very large group (several tens of thousands of individuals) with diverse nutritional needs. To produce enough energy to fly, the foragers themselves need the sugars found in the nectar of plants. The other workers also need energy—primarily to care for the larvae and the nest—but in smaller quantities. In contrast, the larvae require primarily proteins and lipids to grow, just like the queen who must produce and lay eggs, sometimes up to two thousand per day in the case of western honeybees! Audrey Dussutour and her team in Toulouse demonstrated how ant colonies resolve this dietary challenge. The workers in charge of collecting food divide the work of drawing on sources with varying protein and carbohydrate content. They know the colony's needs by analyzing the food reserves in the nest, the possible release of excess food from the nest by other workers, and the composition of the colony. The more larvae the colony contains, the more protein-rich food the ants bring in. Conversely, if the colony is mainly composed of adults, the ants concentrate on gathering carbohydrates. Even more surprisingly, this collective regulation of food intake isn't only observed for the three classes of macronutrients: carbohydrates, lipids, and proteins. Ant colonies are also able to select foods because they contain essential micromolecules, in particular certain amino acids.[22] Insects therefore have the ability to recognize specific nutrients in food independently of other nutrients, which allows them to address potential nutritional deficiencies.

The nutritional stresses experienced by bees are most often the result of landscape modifications from human activities. Put yourself in the shoes of a bee whose habitat is near a monoculture as far as the eye can see, like rapeseed fields in the spring or sunflower fields in the summer. Despite the beauty of these flower-filled expanses and the immense amount of food they provide, these crops greatly reduce the diversity of food available. Due to human intervention, the world is becoming increasingly depleted. While hundreds of plant species once lined our countryside, now only one type of nectar and pollen remains in these fields. Cultivated seeds have been selected according to a set of criteria that allow for a better agricultural yield, but they were never designed to cover the nutritional needs of pollinators. It would be a huge stroke of luck if this single food source could provide a balanced nutritional intake for a colony of bees. Unfortunately, with these monocultures, bees suffer from nutritional deficiencies. Moreover, by flowering only once or twice during the year, these fields, which are full of human food, paradoxically create long periods of scarcity and starve the bees.

Sharoni Shafir and his team at the Hebrew University of Jerusalem tested the significance of omega-3, which honeybees regularly experience in the cultivated landscapes of Israel.[23] These fatty acids are essential for the brain's development and functioning. They're not synthesized by most animals and must therefore be acquired from food. These researchers fed colonies with pollens enriched or depleted in omega-3. As a result, they measured poorer learning and memory performance in bees on the depleted diet. Pollen analysis of plants around the study site revealed low levels of omega-3 in cultivated species such as eucalyptus, apple, pear, and almond trees, revealing the malnutrition that bees endure in these agricultural landscapes. This phenomenon is not specific to the Middle East region and raises the more general issue of single-crop farming. Each year, 60% of the bee colonies raised in the United States are moved to California to pollinate almond trees, creating a major malnutrition problem.

Waiting for the Next World

Bees live in our midst in a polluted and degraded world. While some places on Earth have been more spared than others, our footprint has global effects on the environment. But fortunately, nature is resilient—at least up to a certain point. During the COVID-19 pandemic, the unprecedented pause in human activity on all continents was a breath of fresh air for the planet and its inhabitants. We first saw animals wandering around cities, in places they weren't used to being. Then came the explosion of insect populations everywhere. I'd never observed so many carpenter bees* and wasps as during the calm spring of 2020. I even found myself hosting an orchard bee† on my balcony in the middle of town. In fact, I think that spring was the only time an insect has ever deigned to take up residence in the structure bamboo that had been cobbled together on my balcony a few years earlier, which some people I know referred to as an "insect hotel." As we can see, nature is resilient. But having humans live in permanent lockdown is of course not a solution for stemming the loss of biodiversity. Fleeing to other planets isn't a solution either and would only transfer the problem there. So what can we do?

We often think solutions must come first from governments and corporations because they have the greatest influence on our way of life by setting laws and prices. While this is probably true in the short term, longer-term solutions require a better understanding of the world around us. From some basic research on insect cognition, we can explore sustainable-development strategies that respect the environment.

Research on bee intelligence has revealed their Achilles' heel—that their brains are so small and optimized that even a single neuron being damaged by a stressor can cause cognitive dysfunction. Based on these observations, a simple first step for us to consider would be to significantly reduce the use of insecticides, without necessarily banning them entirely.[24] Currently most of these compounds are used preventively at doses calculated to kill the targeted

* The violet carpenter bee (*Xylocopa violacea*) is a large solitary bee common in Europe. Its body is entirely black with metallic blue reflections, and its wings are dark with purple hues. Its strong mandibles allow it to build its nest in wood, a skill that earned it a fitting name.

† The European orchard bee (*Osmia cornuta*) is a small black solitary bee with a rust-colored abdomen. Mated females build their nests in elongated cavities by forming a series of cells that are separated by clay partitions and that contain food reserves on which the eggs are laid (first the females, then the males). The adults don't leave the nest until one year later.

insects without even verifying that those insects are present in the area being treated. Even at low doses, these compounds spread in the environment and impact non-target species in the form of residues, as shown by the example of neonicotinoids that cause disorientation in bees.* These environmental risks can no longer be justified. And what is true for pollinators is also true for other insects. Therefore a first option in the short term would be to drastically limit the use of insecticides by applying them at low sublethal doses in a targeted manner. This would expose only pests and thus reduce the ecological impact. Another option would be to better control the timing of insecticide use to prevent species from developing resistance as a result of mass treatments. Indeed, the number of resistant species on which insecticides no longer have any effect is constantly increasing, which forces chemists to develop increasingly toxic insecticides. This race to find the right poisons could be avoided by rotating the crops being treated and the insecticides being used. Finally, using fewer insecticides would also decrease costs for farmers, which would consequently increase the net return to the farm. These recommendations don't replace the need to reform intensive agricultural practices in favor of biological and mechanical pest control practices. But they can be applied quickly and would have an immediate positive effect on the environment while not penalizing farmers.

Research on bee intelligence also suggests we should implement new precision agriculture practices. We already see signs of a pollination crisis in the near future due to the ever-increasing demand to feed humans and the declining numbers of pollinators around the world.[25] One strategy to counteract this crisis is to increase bees' natural pollination activity. Walter Farina and his team at the University of Buenos Aires in Argentina extensively studied the role of

* See the section The Harmful Effects of Nicotine in this chapter.

scent learning in western honeybees' waggle-dance recruitment to food sources.* In particular, their research led them to manipulate this type of learning to improve crop pollination.[26] The idea stemmed from the observation that just as a forager learns to recognize floral scents, a worker can learn a flower's scent when in contact with a dancer who smells of the flower it just visited. A single exposure to this scent, even indirect, is sufficient for the future forager bee to form a long-term memory. These researchers demonstrated that when a synthetic floral scent was released into a bee colony, it activated a scent memory and biased their foraging activity toward the flowers in question. By feeding colonies nectar scented with the sunflower mimic odor, the team was able to create olfactory memories in the bees and stimulate sunflower foraging. With this ploy, they were able to increase bee pollination yields by nearly 60% in the fields being tested. Such a yield no longer justifies the massive use of pesticides. For this method to be viable, of course, it's necessary to ensure that bee colonies are in close proximity to wild flora or food supplements to compensate for the possible nutritional imbalance associated with single-crop farming.

Research on bee navigation is another source of simple strategies that could optimize natural pollination processes without the need for chemical treatments. If we can understand how bees move across the landscape and how they choose the flowers they visit, we'll be able to understand the flow of pollen between plants and thus the success of pollination. In fact, we don't really understand why this is, but some plants in a field receive more pollen than others, while other plants have a greater probability of receiving pollen of lower quality or from incompatible plants (because they're of different species, for instance). By anticipating bee behavior, we could directly influence crop pollination, adjusting the number of

* See chapter 1, A Poor Sense of Direction.

pollinators in a given area to levels that ensure all plants are visited. We could also arrange plants in such a way as to encourage pollinators to visit them, for example by placing plants of interest on pollinators' likely flight paths. We should ask ourselves whether a field with parallel rows of plants of the same species is really the best configuration for optimizing pollination by bees, or if other designs or planting a mix of different plant species together should be considered. These strategies can be used to increase agricultural production by promoting polyculture and permaculture methods. This knowledge gleaned from bees also gives us a better understanding of the environmental characteristics that favor foraging and ensure the sustainability of bee populations. These are the questions my team is currently investigating. Several of my colleagues are already working on redesigning environments in areas where pollinators are in critical decline. These new designs would provide bees with easier access to abundant and diverse food sources throughout the year—an approach that could easily be expanded to encourage the establishment and vitality of wild bee communities.

Beyond applying scientific discoveries in a concrete way to change our agricultural practices, a more systematic dissemination of knowledge will make a broader audience aware of the challenges of biodiversity loss and could spark new ideas. As media and researchers are sharing scientific discoveries, things are changing. New ideas are often born from information exchange—and can even come from our youngest citizens. Recently a preschooler asked me how bumblebees learn to fly. What's certain is that in a colony there are always bumblebees that excel at this task and others that never do. But why? I didn't know how to answer the question. But it haunted me for several days, and I now have research plans to attempt to understand it.

The COVID-19 public health crisis showed how limited the communication between researchers and the general public can sometimes be. In a short period of time, people saw numerous experts present scientific knowledge that would then evolve over the course of days, sometimes leading to divergent conclusions, until finally convergent findings from several laboratories would result in a stable scientific consensus. This confusion undermined scientific credibility in the eyes of the public, leading many people to develop their own opinions about the pandemic and the solutions that should be implemented, which were not based on scientific grounds but rather on subjective experience.

A newer way to raise mass awareness is to involve the public directly in the research process. I've been lucky enough to be a part of several of these projects in which volunteers not associated with any research institution collect data and even participate in implementing experimental protocols. For the participants, these projects are an immersive experience in the world of research. For the researchers, they're a chance to answer new questions using huge datasets gathered at a lower cost than usual. My first experience in participatory science came after observing a large presence of invasive Asian hornets (*Vespa velutina*) in our apiary. Since the early 2000s this hornet, which is easily recognized by its yellow legs, has been building huge paper nests containing tens of thousands of workers in our space—and feeding on our bees. Beekeepers try to trap hornets en masse or destroy the nests to protect their hives. But the hornet is nevertheless a fascinating social insect which, like honeybees, lives in a sophisticated society and pollinates in its spare time. It's believed that a queen of this species was unintentionally imported from Southeast Asia to southwest France in 2004. Since then, a population of Asian hornets has invaded much of Europe, from Portugal to Italy, as well as the Netherlands and the United Kingdom. Among these hornets, it's the queens that disperse once a year, sometimes over several dozen miles, to found a new nest each spring. With Antoine Wystrach, a fellow ant specialist who also enjoys sharing coffee in the sun, I hypothesized that these invasive populations were in the process of evolving. If this proved to be the case, the hornets on the front lines of the invasion would need to have morphological traits that favor dispersal, such as large wings to fly farther. On the other hand, the hornets remaining at the introduction site would need to have morphological traits that favor competition with other hornets, for example being larger.

To test our hypothesis, we launched a call for volunteers, asking anyone interested to set a hornet trap for a few days in their yard and send us the captured queens in an envelope in the mail. The operation was an unexpected success. In just a few weeks, we received nearly six thousand hornets from all over Europe, along with jars of honey, nests, and notes of encouragement. After several months of cleaning, sterilizing, dissecting, and measuring these animals, we were able to validate our theory and show that hornet queens (called "foundresses") at the invasion front were evolving almost before our eyes. Their lighter bodies and longer wings increased their flight efficiency and explained the rapid invasion. In addition to the joy of having validated our hypothesis, we had the pleasure of sharing this research with over five hundred participants, without whom the study would never have been conducted. Biological invasions are commonplace, and they teach us that there are unfortunately few actions we can take to combat them. Most often we have to wait for a balance of ecological interactions to be established between these predators and their new prey. I hope that this study helped participants to improve their opinion of these insects. Hornets are often hated by beekeepers, but in reality they're equal to honeybees in terms of their social organization and likely also in terms of their sophisticated individual and collective cognitive abilities.

Sometimes good ideas come from the volunteers themselves. A few years ago, I was contacted by a high-school teacher in the Manche, in Normandy. Three of his students had chosen to work on a sensor system to study bee behavior based on the "connected hive" model. At that time in the beekeeping world, sensors for hives were in the development stage. Professionals and hobbyists alike were creative in equipping their hives with weather stations, scales, and temperature and humidity sensors to monitor the condition of their colonies and their honey production in real time and from a distance. To increase the technical challenge, we had the idea to develop "connected flowers," which would allow us to remotely study the foraging activity of bees in a natural environment. These novel devices recorded the identity of the bees that visited them, as well as the time of their visits. Although this student-initiated work didn't result in a fully finalized product, it proved to be an excellent foundation for subsequent research. Today the project benefits from several million euros in funding and has taken on an academic dimension, involving several laboratories, students from different disciplines, various organizations, local residents, and some companies. The connected flowers have evolved into automated IQ tests to study the cognitive health of bees in different types of environments. We've even extended the concept to other species. As I write this book, connected flowers, bird feeders, and aquariums are being deployed throughout the Toulouse region to measure the health of terrestrial and aquatic ecosystems using model organisms. In this project, western honeybees, great tits, and *Lymnaea* (small aquatic snails) are the "sentinel species." We have good knowledge of these species, making it relatively easy to conduct experiments to measure the effects of environmental changes on them. Based on these model species, much larger groups and the ecological functions they provide can be studied. Strictly speaking, the western honeybee is not in decline. Although the average lifespan of colonies is decreasing, the number of beekeepers is growing, and they're constantly replenishing their bee populations. It's from studying these honeybees' stress response, however, that we can obtain valuable information on the health of wild bee populations, which often suffer significant and alarming declines.

Unlike Achilles and many other heroes of Greek mythology, the fate of bees is not set in stone. With modern research tools and large-scale approaches involving human and sensor networks, we now have all the tools in hand to make a precise diagnosis of the influence that combined environmental stressors exert on animal behavior in the wild and over the long term. If you've read this entire book, you'll no longer wonder why this effort is so important.

EPILOGUE

After All, We're All a Bit Like Bees

One day someone said to me, "It makes sense that you're so interested in bumblebees because you look like one!" I'd already been told this when I was studying cockroaches, and while a weird and slightly offensive thing to say, at the time, I didn't give it much thought. But occasionally I think back on it. What did this comment mean? Was it a form of recognition? (Acknowledging that it's important that people are doing this work.) A polite way to end the conversation? (Meaning that bumblebees are interesting but not for everyone.) Or was it to be taken at face value? (I really do physically resemble a bumblebee.) I'll probably never really know the answers to these questions. (Although arguably I don't have enough hair left on my head to resemble a bee.) But the more I think about the comment, the more I tell myself that deep down we all resemble bumblebees a bit. Let me explain.

Bumblebees and humans share a common ancestor that is estimated to have existed over five hundred million years ago.[1] We also share this ancestor—likely an aquatic worm—with many other animal species. On the one hand, this ancestor led to vertebrates—first fish, then amphibians that colonized the earth and from which reptiles, birds, and mammals descended. On the other hand, this common ancestor with bumblebees led to crustaceans, some of which evolved into insects. Like us, insects have a brain, a heart, and digestive and reproductive organs. They need oxygen and food to live. They get sick and develop immune responses. All these characteristics are written in a genetic code, known as the "genome." At the beginning of the 20th century, the rendering of the human and honeybee genomes uncovered genetic similarities that were previously unimagined to exist. We share over 60% of our genes with bees. Most of our enzymes—the proteins involved in basic physiological functions such as digestion, nerve conduction, and hormone synthesis—are similar. Our nerve cells work the same way. And the similarities don't end there. As we've seen, there is even a certain similarity between our psychologies.

Over the course of evolution, the cognitive abilities that define intelligence appeared, became specialized, and sometimes disappeared and reappeared throughout the animal kingdom. Today it's clear that we share many of these abilities with the animals around us, including those that look the least like us. The more we look for these sophisticated intellectual capacities in insects, the more we find them. Neither the size nor the structure of an animal's brain is indicative of its intellectual capacities.[2] Insects, even with their simple nervous system, show signs of numeracy, attention processes, and other astonishing cognitive abilities, such as individual recognition, object recognition, symbolic communication, tool use, and causal reasoning. Bees in particular seem to have a mental representation of space and time, which is the basis of consciousness. So, contrary to what was thought until about fifty years ago, insects are so much more than reflex mechanisms.

These amazing cognitive similarities between humans and insects naturally lead us to consider insects' physical and mental suffering, or in other words, their well-being. We live in a time when many initiatives are seeking to advance bioethics and animal rights.

Insects, and invertebrates in general, are absent from these discussions, however. Current environmental protection criteria are based on the belief that only vertebrates are "sentient" beings, meaning capable of feeling sensations and emotions, whether positive or negative. The only exceptions are some cephalopods, such as octopus and cuttlefish.* This doesn't mean that people who study insects don't respect them. The vast majority of researchers engage in ethical practices as a result of individual awareness, and they apply the guiding principles of the "three Rs,"† which aim to reduce the number of animals in the laboratory and their potential suffering during experimentation. To date, however, there are no official regulations for the welfare of insects, and many scientists, organizations, and politicians agree that the progress being made around regulations is insufficient. Beyond the annual use of billions of insects for research purposes, we also face the immense challenge of large-scale insect cultivation, which is expected to increase all over the world. Breeding this "mini livestock" is a key element in feeding an ever-growing human population.

As we've seen throughout this book, the exclusion of insects from public debate is no longer supported by scientific knowledge. When we talk about well-being, we often focus on pain, which is a negative emotion perceived by the brain. Therefore, the discovery of emotion-like expressions in bumblebees, honeybees, and *Drosophila* flies is an important argument for permanently changing our perception of these animals. Following the observation of a grasshopper that continued to eat while being eaten itself by a

---

* In Europe, octopus and cuttlefish are the only invertebrates to which ethical laws apply. For more information, see Directive 2010/63/EU dated September 22, 2010.

† These principles, introduced in 1959 by William Stratton Russell and Rex Burch, provide an ethical framework for the use of animals in scientific research according to three main principles: replacement (avoid using animals when possible), reduction (use fewer animals), and refinement (minimize stress and pain).

praying mantis, it was long believed that insects don't feel pain.[3] In all likelihood, this observation was incomplete, and the grasshopper was already dead but appeared to still be moving because its residual nerve activity generated the reflex contraction of muscles. Although research on insects' pain and emotions is recent and still limited, several studies strongly suggest these sensations exist, at least in bees and flies.[4] Since cephalopods are considered "honorary vertebrates," it would be consistent to treat insects the same way.

For these reasons, more and more researchers are proposing the precautionary principle be enforced.[5] History has repeatedly shown us that judging species by their mere appearance doesn't work. For example, birds were long considered to be unintelligent animals with primitive brains, but some people today consider them to be "feathered apes."[6] When an animal exhibits behaviors similar to ours when subjected to the same stimuli, that animal has probably had a comparable internal experience.[7] What if bees are "miniature apes"? The time has come to adopt stricter rules to protect their well-being while we accumulate more data to better understand them.

Despite appearances, we're much more like insects than we might realize. Studying their intelligence is therefore an attempt to better understand our own intelligence, as well as a chance to become aware that we fully belong to the animal world. Personally, I stopped attacking anthills and trapping firebugs in a jar without food a long time ago. Insects have a history going back four hundred million years. The three hundred thousand years that have elapsed since the appearance of the first humans seem quite insignificant by comparison.[8] Let's open our eyes to this world belonging to the miniscule beings whose intelligence has forged unique paths.

ACKNOWLEDGMENTS

I would like to thank Jessica Serra for suggesting I write this book. What I've written in these pages is the accumulation of the work of several generations of passionate and talented researchers, without whom we would still be largely ignorant of the inner life of insects. Some of these researchers, with whom I've had the pleasure of sharing adventures, are named in the book. Others are not, but they'll recognize themselves. This project wouldn't have been possible without the support of my wonderful research group, my family, and my own two little bees.

NOTES

Foreword

1. Ingemar Düring, "Aristote: Histoire des animaux, T.: 1, Livres 1–4," *Gnomon* 37, no. 7 (November 1965): 664–666.
2. Simon Byl, "Aristote et le monde de la ruche," *Revue belge de philologie et d'histoire* 56, no. 1 (1978): 15–28.
3. Pliny the Elder, *The Natural History*, trans. John Bostock (London: Taylor and Francis, 1855).
4. Gregory of Nyssa, Homily 9, *Homilies on the Song of Songs*, trans. Richard A. Norris Jr. (Atlanta: Society of Biblical Literature, 2012).
5. Pierre-Henri Tavoillot and François Tavoillot, *L'abeille (et le) philosophe: étonnant voyage dans la ruche des sages* (Paris: Odile Jacob, 2015).
6. "L'abeille ou le miroir des idées," *L'Humanité*, September 7, 2015.

Prologue. What Is a Bee?

1. René Descartes, *Discours de la méthode* (1637).
2. Karl von Frisch, *Aus dem Leben der Bienen* (Berlin: Springer-Verlag, 1927); Thomas D. Seeley, *Honeybee Democracy* (Princeton, NJ: Princeton University Press, 2010); Dave Goulson, *A Sting in the Tale: My Adventures with Bumblebees* (London: Jonathan Cape, 2013).
3. Henri Clément, *Le traité rustica de l'apiculture* (Paris: Rustica Éditions, 2018).
4. Jakob von Uexküll, *Streifzüge durch die Umwelten von Tieren und Menschen* (Berlin: Verlag von Julius Springer, 1934); *Bedeutungslehre* (Leipzig: Verlag von J. A. Barth, 1940).

1. A Poor Sense of Direction

1. Randolf Menzel and Martin Giurfa, "Cognitive architecture of a minibrain: The honeybee," *Trends in Cognitive Sciences* 5, no. 2 (February 1, 2001): 62–71.

2. Dominic Clarke et al., "Detection and learning of floral electric fields by bumblebees," *Science* 340, no. 6128 (April 5, 2013): 66–69.
3. Mario Pahl et al., "Large scale homing in honeybees," *PLOS ONE* 6, no. 5 (May 13, 2011): e19669.
4. Bernd Heinrich, *Bumblebee Economics* (Cambridge: Harvard University Press, 2004).
5. Karl von Frisch, *Tanzsprache und Orientierung der Bienen* (Berlin: Springer-Verlag, 1965).
6. Harald Esch and John Burns, "Distance estimation by foraging honey-bees," *Journal of Experimental Biology* 199, no. 1 (January 1996): 155–162.
7. Mandyam V. Srinivasan et al., "Honeybee navigation: Nature and calibration of the 'odometer,'" *Science* 287, no. 5454 (February 4, 2000): 851–853.
8. Matthias Wittlinger, Rüdiger Wehner, and Harald Wolf, "The ant odometer: Stepping on stilts and stumps," *Science* 312, no. 5782 (June 30, 2006): 1965–1967.
9. Von Frisch, *Tanzsprache und Orientierung der Bienen*.
10. James F. Cheeseman et al., "Way-finding in displaced clock-shifted bees proves bees use a cognitive map," *Proceedings of the National Academy of Sciences of the United States of America* 111, no. 24 (June 2, 2014): 8949–8954.
11. R. B. Freeman, "Charles Darwin on the routes of male humble bees," *Bulletin of the British Museum (Natural History) Historical Series* 3, no. 6 (1968): 177–189.
12. Mathieu Lihoreau, Lars Chittka, and Nigel E. Raine, "Travel optimization by foraging bumblebees through readjustments of traplines after discovery of new feeding locations," *American Naturalist* 176, no. 6 (December 2010): 744–757.
13. *Guardian*, "Bees' tiny brains beat computers, study finds," October 24, 2010.
14. Mathieu Lihoreau et al., "Radar tracking and motion-sensitive cameras on flowers reveal the development of pollinator multi-destination routes over large spatial scales," *PLOS Biology* 10, no. 9 (September 20, 2012): e1001392.

15. Randolf Menzel et al., "A common frame of reference for learned and communicated vectors in honeybee navigation," *Current Biology* 21, no. 8 (April 26, 2011): 645–650.
16. J. O'Keefe and J. Dostrovsky, "The hippocampus as a spatial map: Preliminary evidence from unit activity in the freely-moving rat," *Brain Research* 34, no. 1 (November 12, 1971): 171–175.
17. Joanna Brebner et al., "Bumble bees strategically use ground level linear features in navigation," *Animal Behaviour* 179, no. 7 (September 2021): 147–160.
18. Elisa Frasnelli, Natalie Hempel de Ibarra, and Finlay J. Stewart, "The dominant role of visual motion cues in bumblebee flight control revealed through virtual reality," *Frontiers in Physiology* 9, no. 1038 (July 31, 2018): 1038.

2. The Fragrance of Déjà Vu

1. Edward O. Wilson, *Sociobiology: The New Synthesis* (Cambridge: Harvard University Press, 1975).
2. Vanina Vergoz, Haley A. Schreurs, and Alison R. Mercer, "Queen pheromone blocks aversive learning in young worker bees," *Science* 317, no. 5836 (July 20, 2007): 384–386.
3. Annette Van Oystaeyen et al., "Conserved class of queen pheromones stops social insect workers from reproducing," *Science* 343, no. 6168 (January 17, 2014): 287–290.
4. Patrizia d'Ettore et al., "Wax combs mediate nestmate recognition by guard honeybees," *Animal Behaviour* 71, no. 4 (April 2006): 773–779.
5. K. J. Pfeiffer and K. Crailsheim, "Drifting of honeybees," *Insectes Sociaux* 45, (1998): 151–167.
6. Tatiana Giraud, Jes S. Pedersen, and Laurent Keller, "Evolution of supercolonies: The Argentine ants of southern Europe," *Proceedings of the National Academy of Sciences of the United States of America* 99, no. 9 (April 16, 2002): 6075–6079.
7. W. D. Hamilton, "The genetical evolution of social behvaiour I & II," *Journal of Theoretical Biology* 7, no. 1 (July 1964): 1–52.

8. William J. Bell, Louis M. Roth, and Christine A. Nalepa, *Cockroaches: Ecology, Behavior, and Natural History* (Baltimore: Johns Hopkins University Press, 2007).
9. Mathieu Lihoreau, Cédric Zimmer, and Colette Rivault, "Kin recognition and incest avoidance in a group-living insect," *Behavioral Ecology* 18, no. 5 (September 2007): 880–887.
10. Mathieu Lihoreau, Colette Rivault, and Jelle S. van Zweden, "Kin discrimination increases with odor distance in the German cockroach," *Behavioral Ecology* 27, no. 6 (November–December 2016): 1694–1701.
11. Elizabeth A. Tibbetts, "Visual signals of individual identity in the wasp *Polistes fuscatus*," *Proceedings of the Royal Society B: Biological Sciences* 269, no. 1499 (July 22, 2002): 1423–1428.
12. Kevin N. Laland, "Animal cultures," *Current Biology* 18, no. 9 (May 6, 2008): R366–R370.
13. Ellouise Leadbeater and Lars Chittka, "A new mode of information transfer in foraging bumblebees?," *Current Biology* 15, no. 12 (June 21, 2005): R447–R448.
14. Erika H. Dawson et al., "Learning by observation emerges from simple associations in an insect model," *Current Biology* 23, no. 8 (April 22, 2013): 727–730.
15. Sylvain Alem et al., "Associative mechanisms allow for social learning and cultural transmission of string pulling in an insect," *PLOS Biology* 14, no. 10 (October 4, 2016): e1002564.
16. Olli J. Loukola et al., "Bumblebees show cognitive flexibility by improving on an observed complex behavior," *Science* 355, no. 6327 (February 24, 2017): 833–836.
17. Étienne Danchin et al., "Cultural flies: Conformist social learning in fruitflies predicts long-lasting mate-choice traditions," *Science* 362, no. 6418 (November 30, 2018): 1025–1030.
18. Danchin et al., 1025–1030.
19. Christopher Krupenye and Josep Call, "Theory of mind in animals: Current and future directions," *WIREs Cognitive Science* 10, no. 6 (November/December 2019): e1503.

20. David Premack and Guy Woodruff, "Does the chimpanzee have a theory of mind?," *Behavioral and Brain Sciences* 1, no. 4 (December 1978): 515–526.
21. Elizabeth A. Tibbetts, Ellery Wong, and Sarah Bonello, "Wasps use social eavesdropping to learn about individual rivals," *Current Biology* 30, no. 15 (August 3, 2020): 3007–3010.

3. The Limits of a Miniature Intelligence

1. Martin Hammer, "An identified neuron mediates the unconditioned stimulus in associative olfactory learning in honeybees," *Nature* 366 (November 4, 1993): 59–63.
2. Frederic Libersat and Ram Gal, "Wasp voodoo rituals, venom-cocktails, and the zombification of cockroach hosts," *Integrative and Comparative Biology* 54, no. 2 (July 2014): 129–142.
3. Mario Pahl et al., "Circadian timed episodic-like memory—A bee knows what to do when, and also where," *Journal of Experimental Biology* 210, no. 20 (October 15, 2007): 3559–3567.
4. John Lubbock, "Observations on ants, bees, and wasps—Part X," *Journal of the Linnean Society of London, Zoology* 17, no. 98 (April 1883): 41–52.
5. Karl von Frisch, "Der Farbensinn und Formensinn der Biene," *Zoologische Jahrbücher* 35 (1915): 1–188.
6. Hiruni Samadi Galpayage Dona and Lars Chittka, "Charles H. Turner, pioneer in animal cognition," *Science* 370, no. 6516 (October 30, 2020): 530–531.
7. C. H. Turner, "The homing of the burrowing-bees (Anthophoridae)," *Biological Bulletin* 15, no. 6 (November 1908): 247–258.
8. C. H. Turner, "Experiments on color-vision of the honey bee," *Biological Bulletin* 19, no. 5 (October 1910): 257–279.
9. Martin Giurfa and Maria Gabriela de Brito Sanchez, "Black Lives Matter: Revisiting Charles Henry Turner's experiments on honey bee color vision," *Current Biology* 30, no. 20 (October 19, 2020): R1235–R1239.
10. Martin Giurfa and Josué A. Núñez, "Honeybees mark with scent and reject recently visited flowers," *Oecologia* 89 (January 1992): 113–117.

11. Kimihisa Takeda, "Classical conditioned response in the honey bee," *Journal of Insect Physiology* 6, no. 3 (July 1961): 168–179.
12. Takeda, 168–179.
13. Maria Eugenia Villar et al., "Refining single-trial memories in the honeybee," *Cell Reports* 30, no. 8 (February 25, 2020): 2603–2613.
14. Jasdan Joerges et al., "Representations of odours and odour mixtures visualized in the honeybee brain," *Nature* 387 (May 15, 1997): 285–288.
15. Jean-Marc Devaud et al., "Neural substrate for higher-order learning in an insect: Mushroom bodies are necessary for configural discriminations," *Proceedings of the National Academy of Sciences of the United States of America* 112, no. 43 (October 12, 2015): 5854–5862.
16. Eleanor A. Maguire et al., "Navigation-related structural change in the hippocampi of cab drivers," *Proceedings of the National Academy of Sciences of the United States of America* 97, no. 8 (March 14, 2000): 4398–4403.
17. Martin Giurfa et al., "The concepts of 'sameness' and 'difference' in an insect," *Nature* 410 (April 19, 2001): 930–933.
18. Scarlett R. Howard et al., "Numerical ordering of zero in honey bees," *Science* 360, no. 6393 (June 8, 2018): 1124–1126.
19. Lars Chittka and Karl Geiger, "Can honey bees count landmarks?," *Animal Behaviour* 49, no. 1 (January 1995): 159–164.
20. Julie Bernard and Martin Giurfa, "A test of transitive inferences in free-flying honeybees: Unsuccessful performance due to memory constraints," *Learning & Memory* 11, no. 3 (May 2004): 328–336.
21. Elizabeth A. Tibbetts et al., "Transitive inference in *Polistes* paper wasps," *Biology Letters* 15, no. 5 (May 2019): 20190015.
22. Mathieu Lihoreau and Colette Rivault, "Tactile stimuli trigger group effects in cockroach aggregations," *Animal Behaviour* 75, no. 6 (June 2008): 1965–1972.
23. Clint J. Perry, Luigi Baciadonna, and Lars Chittka, "Unexpected rewards induce dopamine-dependent positive emotion-like state changes in bumblebees," *Science* 353, no. 6307 (September 30, 2016): 1529–1531.
24. Perry, Baciadonna, and Chittka, 1529–1531.

25. David Baracchi et al., "Nicotine in floral nectar pharmacologically influences bumblebee learning of floral features," *Scientific Reports* 7 (May 16, 2017): 1–8.
26. Galit Shohat-Ophir et al., "Sexual deprivation increases ethanol intake in *Drosophila*," *Science* 335, no. 6074 (March 16, 2012): 1351–1355.
27. Cwyn Solvi, Selene Gutierrez Al-Khudhairy, and Lars Chittka, "Bumble bees display cross-modal object recognition between visual and tactile senses," *Science* 367, no. 6480 (February 21, 2020): 910–912.
28. Sridhar Ravi et al., "Bumblebees perceive the spatial layout of their environment in relation to their body size and form to minimize inflight collisions," *Proceedings of the National Academy of Sciences of the United States of America* 117, no. 49 (November 23, 2020): 31494–31499.
29. Clint Perry and Andrew B. Barron, "Honey bees selectively avoid difficult choices," *Proceedings of the National Academy of Sciences of the United States of America* 110, no. 47 (November 4, 2013): 19155–19159.
30. Michael L. Smith, Nils Napp, and Kirstin H. Peterson, "Imperfect comb construction reveals the architectural abilities of honeybees," *Proceedings of the National Academy of Sciences of the United States of America* 118, no. 31 (July 26, 2021): e2103605118.
31. Vince Gallo and Lars Chittka, "Stigmergy versus behavioral flexibility and planning in honeybee comb construction," *Proceedings of the National Academy of Sciences of the United States of America* 118 (August 12, 2021): e2111310118.
32. Lubbock, "Observations on ants, bees, and wasps—Part X," 41–52.
33. Marie-Geneviève Guiraud, Mark Roper, and Lars Chittka, "High-speed videography reveals how honeybees can turn a spatial concept learning task into a simple discrimination task by stereotyped flight movements and sequential inspection of pattern elements," *Frontiers in Psychology* 9 (August 3, 2018): 1347.
34. Olli J. Loukola et al., "Bumblebees show cognitive flexibility by improving on an observed complex behavior," *Science* 355, no. 6327 (February 24, 2017): 833–836.
35. Lars Chittka and Catherine Wilson, "Expanding consciousness," *American Scientist* 107, no. 6 (November–December 2019): 364–369.

36. Lucy A. L. Tainton-Heap et al., "A paradoxical kind of sleep in *Drosophila melanogaster*," *Current Biology* 31, no. 3 (February 8, 2021): 578–590.
37. Andrew B. Barron and Colin Klein, "What insects can tell us about the origins of consciousness," *Proceedings of the National Academy of Sciences of the United States of America* 113, no. 18 (April 18, 2016): 4900–4908.

4. The Superorganism

1. Paul Tenczar et al., "Automated monitoring reveals extreme interindividual variation and plasticity in honeybee foraging activity levels," *Animal Behaviour* 95 (September 2014): 41–48.
2. William Morton Wheeler, "The ant-colony as an organism," *Journal of Morphology* 22, no. 2 (June 1911): 307–325.
3. Adriano Pimentel Farias et al., "Nest architecture and colony growth of *Atta bisphaerica* grass-cutting ants," *Insects* 11, no. 11 (October 29, 2020): 741.
4. Martin Lindauer, "Bienentänze in der Schwarmtraube," *Naturwissenschaften* 38 (1951): 509–513.
5. Thomas D. Seeley et al., "Stop signals provide cross inhibition in collective decision-making by honeybee swarms," *Science* 335, no. 6064 (December 8, 2011): 108–111.
6. S. Goss et al., "Self-organized shortcuts in the Argentine ant," *Naturwissenschaften* 76, no. 12 (December 1, 1989): 579–581.
7. R. Beckers et al., "Collective decision making through food recruitment," *Insectes Sociaux* 37 (September 1, 1990): 258–267.
8. Takao Sasaki and Stephen C. Pratt, "Groups have a larger cognitive capacity than individuals," *Current Biology* 22, no. 19 (October 9, 2012): R827–R829.
9. Takao Sasaki et al., "Ant colonies outperform individuals when a sensory discrimination task is difficult but not when it is easy," *Proceedings of the National Academy of Sciences of the United States of America* 110, no. 34 (August 20, 2013): 13769–13773.
10. Mathieu Lihoreau et al., "Collective selection of food patches in *Drosophila*," *Journal of Experimental Biology* 219, no. 5 (March 1, 2016): 668–675.
11. T. Pankiw and R. E. Page Jr., "The effect of genotype, age, sex, and caste on response thresholds to sucrose and foraging behavior of honey bees

(*Apis mellifera* L.)," *Journal of Comparative Physiology A* 185, no. 2 (1999): 207–213.

12. Simon Klein et al., "Inter-individual variability in the foraging behaviour of traplining bumblebees," *Scientific Reports* 7, no. 1 (July 4, 2017): 4561.
13. Lars Chittka et al., "Bees trade off foraging speed for accuracy," *Nature* 424, no. 6947 (July 24, 2003): 388.
14. James G. Burns and Adrian G. Dyer, "Diversity of speed-accuracy strategies benefits social insects," *Current Biology* 18, no. 20 (October 28, 2008): R953–R954.
15. A. Dussutour et al., "Individual differences influence collective behaviour in social caterpillars," *Animal Behaviour* 76, no. 1 (July 2008): 5–16.
16. Michael J. B. Krieger, Jean-Bernard Billeter, and Laurent Keller, "Ant-like task allocation and recruitment in cooperative robots," *Nature* 406, no. 6799 (August 31, 2000): 992–995.
17. J. Halloy et al., "Social integration of robots into groups of cockroaches to control self-organized choices," *Science* 318, no. 5853 (November 16, 2007): 1155–1158.
18. Tim Landgraf et al., "Dancing honey bee robot elicits dance-following and recruits foragers," *arXiv* (March 19, 2018): 1803.07126.
19. Franck Bonnet et al., "Robots mediating interactions between animals for interspecies collective behaviors," *Science Robotics* 4, no. 28 (March 13, 2019): eaau7897.
20. E. Bonabeau, M. Dorigo, and G. Theraulaz, "Inspiration for optimization from social insect behaviour," *Nature* 406, no. 6791 (July 6, 2000): 39–42.
21. Thomas D. Seeley, *Honeybee Democracy* (Princeton, NJ: Princeton University Press, 2010).
22. Joseph B. Bak-Coleman et al., "Stewardship of global collective behavior," *Proceedings of the National Academy of Sciences of the United States of America* 118, no. 27 (June 21, 2021): e2025764118.

5. Achilles' Tarsus

1. Michael S. Engel and David A. Grimaldi, "New light shed on the oldest insect," *Nature* 427, no. 6975 (February 12, 2004): 627–630.

2. Bert Hölldobler and Edward O. Wilson, *The Superorganism: The Beauty, Elegance, and Strangeness of Insect Societies* (New York: W. W. Norton & Co, 2009).
3. Francisco Sánchez-Bayo and Kris A. G. Wyckhuys, "Worldwide decline of the entomofauna: A review of its drivers," *Biological Conservation* 232 (April 2019): 8–27.
4. Roel van Klink et al., "Meta-analysis reveals declines in terrestrial but increases in freshwater insect abundances," *Science* 368, no. 6489 (April 24, 2020): 417–420.
5. Simon Klein et al., "Why bees are so vulnerable to environmental stressors," *Trends in Ecology & Evolution* 32, no. 4 (April 2017): 268–278.
6. Mickaël Henry et al., "A common pesticide decreases foraging success and survival in honey bees," *Science* 336, no. 6079 (March 29, 2012): 348–350.
7. James D. Crall et al., "Neonicotinoid exposure disrupts bumblebee nest behavior, social networks, and thermoregulation," *Science* 362, no. 6415 (November 9, 2018): 683–686.
8. Sébastien C. Kessler et al., "Bees prefer foods containing neonicotinoid pesticides," *Nature* 521 (April 22, 2015): 74–76.
9. Harry Siviter, Mark J. Brown, and Ellouise Leadbeater, "Sulfoxaflor exposure reduces bumblebee reproductive success," *Nature* 561 (August 15, 2018): 109–112.
10. Coline Monchanin et al., "Current permissible levels of metal pollutants harm terrestrial invertebrates," *Science of the Total Environment* 779 (July 20, 2021): 146398.
11. Fleur Ponton et al., "Parasites survive predation on its host," *Nature* 440 (April 5, 2006): 756.
12. Clint J. Perry et al., "Rapid behavioral maturation accelerates failure of stressed honey bee colonies," *Proceedings of the National Academy of Sciences of the United States of America* 112, no. 11 (February 9, 2015): 3427–3432.
13. Alessandra Mura et al., "Propolis consumption reduces *Nosema ceranae* infection of European honey bees (*Apis mellifera*)," *Insects* 11, no. 2 (February 15, 2020): 124.

14. Masato Ono et al., "Unusual thermal defence by a honeybee against mass attack by hornets," *Nature* 377 (September 28, 1995): 334–336.
15. Nathalie Stroeymeyt et al., "Social network plasticity decreases disease transmission in a eusocial insect," *Science* 362, no. 6417 (November 23, 2018): 941–945.
16. Livia H. Morais, Henry L. Schreiber IV, and Sarkis K. Mazmanian, "The gut microbiota–brain axis in behaviour and brain disorders," *Nature Reviews Microbiology* 19 (October 22, 2020): 241–255.
17. Cassondra L. Vernier et al., "The gut microbiome defines social group membership in honey bee colonies," *Science Advances* 6, no. 42 (October 14, 2020): eabd3431.
18. Gil Sharon et al., "Commensal bacteria play a role in mating preference of *Drosophila melanogaster*," *Proceedings of the National Academy of the United States of America* 107, no. 46 (November 1, 2010): 20051–20056.
19. Adam Chun-Nin Wong et al., "Gut microbiota modifies olfactory-guided microbial preferences and foraging decisions in *Drosophila*," *Current Biology* 27, no. 15 (August 7, 2017): 2397–2404.
20. Michael DeNieu, Kristin Mounts, and Mollie Manier, "Two gut microbes are necessary and sufficient for normal cognition in *Drosophila melanogaster*," *bioRxiv* (March 30, 2019).
21. D. Raubenheimer and S. J. Simpson, "The geometry of compensatory feeding in the locust," *Animal Behaviour* 45, no. 5 (May 1993): 953–964.
22. Enikő Csata et al., "Ant foragers compensate for the nutritional deficiencies in the colony," *Current Biology* 30, no. 1 (January 6, 2020): 135–142.
23. Yael Arien et al., "Omega-3 deficiency impairs honey bee learning," *Proceedings of the National Academy of Sciences of the United States of America* 112, no. 51 (December 7, 2015): 15761–15766.
24. Constance Holden, "Report warns of looming pollination crisis in North America," *Science* 314, no. 5798 (October 20, 2006): 397.
25. Walter M. Farina et al., "Learning of a mimic odor within beehives improves pollination service efficiency in a commercial crop," *Current Biology* 30, no. 21 (November 2, 2020): 4284–4290.
26. Théotime Colin et al., "Pesticide dosing must be guided by ecological principles," *Nature Ecology & Evolution* 4 (September 7, 2020): 1575–1577.

Epilogue. After All, We're All a Bit Like Bees

1. Stephen Jay Gould, *Wonderful Life* (New York, W. W. Norton & Co, 1989).
2. Lars Chittka and Jeremy Niven, "Are bigger brains better?," *Current Biology* 19, no. 21 (November 17, 2009): R995–R1008.
3. C. H. Eisemann et al., "Do insects feel pain? A biological view," *Experientia* 40 (February 1984): 164–167.
4. David Baracchi and Luigi Baciadonna, "Insect sentience and the rise of a new inclusive ethics," *Animal Sentience* 29, no. 18 (2020).
5. Claudio Carere and Jennifer Mather, eds., *The Welfare of Invertebrate Animals* (New York: Springer, 2019).
6. Nicola Clayton, "Corvid cognition: Feathered apes," *Nature* 484 (April 25, 2012): 453–454.
7. A. van Huis, "Welfare of farmed insects," *Journal of Insects as Food and Feed* 5, no. 3 (July 9, 2019): 159–162.
8. Jean-Jacques Hublin et al., "New fossils from Jebel Irhoud, Morocco, and the pan-African origin of *Homo sapiens*," *Nature* 546 (June 8, 2017): 289–292.